BRITISH RAILW

STANDARD

STEAM LOCOMO

BRITISH RAILWAYS STANDARD STEAM LOCOMOTIVES

E. S. Cox
M.I.Mech.E., M.I.Loco.E.

LONDON
IAN ALLAN

First published 1966

This edition 1973

ISBN 0 7110 0449 8

Published by Ian Allan Ltd., Shepperton, Surrey
and printed in the United Kingdom by
The Press at Coombelands Ltd., Addlestone, Weybridge, Surrey

CONTENTS

Unless otherwise stated all photographs in this book are by courtsey of British Railways.

LIST OF TABLES

PREFACE

IT WAS that most experienced literary craftsman Somerset Maugham who said that while the well-known saying asserts that everyone has it in him to write one book, it says nothing about a second. He advises that the amateur in writing is wise not to try his luck again for what he produces next is pretty sure to be worthless. Compared to the 48 years span of my first book *Locomotive Panorama*, only 12 short years cover the whole design and production of the British Railways standard locomotives with which the present volume deals. Less than one thousand of them were built as against the more than eleven thousand which were previously described. Thus even if not entirely worthless, it is clearly more difficult to make the second book as interesting as the first.

On the other hand, the incidence of the B.R. locomotives was an adventurous rearguard action as the steam era drew to its close. It was a defiant gesture against those who on the one hand considered it impious to tamper with the final productions of the great masters of steam, and against those on the other hand who thought that the walls of Jericho should fall down flat and at once in the face of the still distant trumpet blasts of the approaching new forms of motive power.

Enthusiasts and devotees of steam should be glad that there were men of spirit, unwilling to be hypnotised by the oncoming monsters, and who believed they could serve their railway best by bringing steam to a further, even if final, stage of development. After all, nobody knew in 1948, or could even imagine that steam was to be discarded at so frantic a pace as subsequently occurred, and that a reliability and a profitability vacuum was to be thus produced until the diesel and the electric had won sufficient time for their proper development.

So for those who have known and loved the steam locomotive there is perhaps a little more to be said, capable of bearing a certain measure of interest.

In 1960 there was turned out of Swindon Works, a 2–10–0 locomotive No. 92220, which was at one and the same time the last steam engine constructed for British Railways and the final unit of 999 locomotives to British Railways standard designs which had been built since nationalisation. By this act, accompanied by appropriate ceremony, was terminated a century and a half of steam traction development in this country since Trevethick's tiny machine with its vast flywheel first crept along its Welsh track; by the same token was ended the controversial series of locomotives which succeeded the products of the four former main line companies, each rooted in an aura of past traditions and famous names.

In my previous book I devoted a chapter to the B.R. standard steam engines with the inception and development of which I was very closely associated. At Ian Allan's suggestion, I have now expanded this to set out as much further information about them as is available against a background of rapidly vanishing records.

They were not of course the only series of standard locomotives to have been evolved in this country, whatever meaning one cares to attach to the word "standard". Far from it, indeed, and a survey of what had previously been done in this direction seemed to me essential as an introductory chapter.

In describing the origin, design, construction, performance and efficiency of the B.R. standards, it is not only one further such series that is being dealt with, but on a rather larger canvas, it is the final development of the Stephensonian locomotive conception which is being portrayed, a conception so fundamentally right that not all the technical developments of steam during its whole lifetime were able to replace it either here or abroad as the form of the steam machine most apt for rail traction. In the United States, Europe, Africa and Asia, as well as in this country, the last steam engines on rail were no more than enlarged and perfected "Rockets" only superseded in the end by quite other forms of prime mover.

I said above that the B.R. locomotives were controversial, a term with which most locomotive enthusiasts, both professional and lay, would heartily agree. They were controversial firstly because they were a manifestation of nationalisation, of state control, of socialism in action, and as such, open to all kinds of political emotion. They were intended to be a synthesis of best former practices, and so were by definition an affront to the spirit of rugged individualism so deeply entrenched in the British mind. They brought to an end streams of locomotive appearance and character stemming from the great names of the near and distant past, Churchward and Collett, Ivatt the father and Gresley, Stanier and Ivatt the son, Maunsell and Bulleid. They were thought to be the product of a faceless technical civil service as it were, amorphous and without character.

In the last of these counts at any rate, public estimation was at fault. Notwithstanding the fact that every locomotive is the product of countless minds and hands, and that it is bound to owe much to past practice, the B.R. series was primarily the product of a triumvirate—Riddles, Bond and Cox—three men who, sharing a common engineering inheritance and experience, combined to work together as one in the conception, design and production of the final steam locomotive types to be placed upon British rails. What follows is an attempt, however unworthy, to portray the manner in which they worked, and the nature of their product.

I am indebted to J. F. Harrison my former chief, for permission to root out old data from his design offices, and to many erstwhile colleagues and friends, both inside and outside the Railway Service, for help in assembling information and looking up drawings. To R. C. Bond,

Chief Mechanical Engineer and later General Manager Workshops, B.R.B., I again owe sincerest thanks for reading the manuscript and for his helpful contribution. With suitable acknowledgements I have plundered the published articles of C. J. Allen and O. S. Nock seeking the records of day-to-day performance in service which I have never been able to take for myself. Finally I would pay a tribute to R. A. Riddles under whose direction all that is described in these pages took place. I owe him a debt of gratitude for his leadership, encouragement and human tolerance, which made the formative five years from 1948 to 1953 the most active and enjoyable of my whole career.

E. S. Cox

I

Standard Locomotives in Great Britain before 1948

"STANDARD" LOCOMOTIVES existed in Great Britain long before nationalisation and the British Railways standard locomotives which are the main subject of this book. Not only were standard series evolved on the old individual railways, but standardisation was taken seriously on each of the four large groups after 1923, although to a differing extent. Moreover, there have been locomotive series designed and built with the intent or at any rate in knowledge that they would be allocated for duty on some or even all of the currently existing railways. Before the B.R. locomotives can be properly placed in their niche in motive power history, therefore, it is desirable to take a backward glance, and in this the first necessity is to come to grips with the word "Standard". In the passage of time and with accumulating practice, it has come to mean many things to many men. It can for instance mean a group of locomotive types, all different, which has been chosen from a number of others to deal with the various kinds of traffic which offers on a given railway at a given time. Although loosely called standardisation, this is really only limitation of variety of which a good example is the current British Railways series of five so-called standard main line diesel locomotives, each entirely different from the others except in a few details, but nevertheless selected for exclusive current building.

A second meaning implies a series of locomotives to cover the traffic requirements of a given railway, having a number of identical component parts applied to the whole stud. This is true technical standardisation and in the best developed steam examples it included boilers, cylinders, valve gear, wheels, axles, axleboxes, and a multiplicity of fittings and details down to pins, rivets, nuts and bolts.

A third interpretation is where a number of locomotive types, conforming to either the first or second definitions above, or to some combination of both, is designed or chosen for use on more than one railway usually following amalgamations of one kind or another, such as took place in this country in 1923 and again in 1948.

The purpose of standardisation, in whatever form, is an economic one. Briefly it is adopted to achieve some or all of the following:

(a) A reduction in first cost due to quantity production, either a lot of the same kind of locomotive, or several types of locomotives having common components which can be manufactured in bulk.

(b) Reduction in cost of repairs, again by reason of dealing with large quantities of the same thing.

(c) Reduction in time of repairs. The most obvious example here is the adoption of a standard boiler covering several types of engine which can be repaired independently and held in stock, so that a locomotive coming in for repairs can have its own boiler taken off and a spare repaired one put on within the seven or eight days which is all the chassis part demands for its own repair. Otherwise, if its own boiler had to be repaired and put back three to four times the number of days out of traffic would, and in earlier days often did, result.

(d) Reduction in stores stock of spares required to service the repairs of the total stud of locomotives.

(e) Easement of the human activities of driving, firing, servicing, maintaining the locomotive, and training the personnel where common assemblies, components, and details face the driver, fireman or fitter in performing his daily tasks.

These factors are both self obvious and desirable, which is why so many chief mechanical engineers through history have either introduced standard designs of their own, or have perpetuated and developed the standard classes of their predecessors. Few indeed have been those who did not believe in standardisation at all, and who designed and built their engines in small batches all more or less different one from the other. Craven of the L.B.S.C. Railway from 1847-1869, and Fletcher who was in charge of the N.E. Railway locomotive department from 1854-1883, come to mind. In more modern times there is the case of the Caledonian Railway whose 65 4–6–0 locomotives were divided into 11 variants, only two of which were even approximately similar; also of Oliver Bulleid, last C.M.E. of the Southern Railway who sincerely felt that standardisation between types was the enemy of all true progress in locomotive matters, and publicly said so.

On some of the old railways, standardisation was more apparent than real. Different classes of engine could share the same family appearance, and tables of dimensions seemed to indicate common boilers serving several types. This was not so, however, because where, as was often the case, an angle ring joined the boiler barrel to the front tube plate, the lower end of the latter was dimensioned and shaped so that it could be bolted directly on to the top rear flange of the inside cylinders. The cylinders on the other hand could be differently inclined or at a different height below the boiler centre line according to the diameter of the coupled wheels on different classes, with the result, for example, that the otherwise identically dimensioned L.N.W. "Experiment" boiler could not be interchanged with that of the "19-in." 4–6–0 goods engine because the front tubeplates were different. This difficulty was overcome on certain other railways such as the L.Y.R. by the use of the "drumhead" tubeplate inset within the first boiler barrel ring and a separate smokebox saddle was designed to acommodate all the cylinder variations and thus permit ready interchange of boilers. Even with this point safeguarded, there could be another barrier due to differences in the position of the expansion angle irons riveted to the firebox sides which supported the boiler at the back end. Incredible as it may seem, the frames of successive locomotives intended to carry the same boiler, were sometimes designed without regard to this need for identity of expansion angle position on the firebox, with the result that the main benefit of boiler standardisation was lost. I have before me as I write, a document dated 2/4/25 titled "Co-ordination of Design—Boiler Dimensions of L.M.S. Locomotives". In this are tabulated all of the dimensions affecting interchangeability of all the boilers and engines running at that time, and it is an eye-opener to study the gulf which sometimes existed between intention and achievement in this respect.

Another vagary was where a series of locomotives could be built new to strict standards as regards moving parts. If engineering management did not have available fixed limits of wear, and specify progressive stepped sizes to which wearing parts should be repaired, and if it did not enforce these with an iron discipline, then individual engines within the class could quickly become non-standard after successive repairs. Without such overall guidance, each foreman or chargehand would use his own judgement as to how much required taking off or opening out to make good wear. Old timers indeed honestly thought that they were being most economical in only machining away the very minimum amount on pin or hole to eliminate uneven wear, but the result was that each individual engine quickly became unlike any other, and parts from one could not after reconditioning be fitted to any other. Holcroft* describes how, when the Ashford men in command of mechanical matters in the newly formed Southern Railway in 1923, took over affairs at Brighton, they found that for the reasons indicated above, it could not be assumed that any two engines of a class were strictly alike.

As with all things in this life, there is of course another side to standardisation. Intended to be the servant of economic railway operation, it can all too readily become a despotic master inhibiting change. The evils of standardisation are the vested interests which are created towards carrying them on too long, even after they have become out of date, and less effective and economic than what has subsequently become available.

In another place I have indicated the disadvantages under which the later L.N.W.R. engines laboured due to adherence to a design philosophy stretching unbroken from the days of Ramsbottom and Webb. Similarly for all their other excellencies, the G.W.R. persisted in some features of design long overdue for replacement which did not stand up to the test of wider application under Stanier on the L.M.S. and later under nationalisation.

When practices and standards have been running for a generation or more, it requires extraordinary courage or self confidence or both to throw them away and start again. Churchward was pre-eminent in this, and

* *Locomotive Adventure*, Vol. 1, H. Holcroft, Pub. Ian Allan.

he is rightly revered for his conceptions of a completely new motive power series highly standardised within itself, owing practically nothing to previous practice, and brought to virtual completion during his own term of office. Aspinall did the same thing for the L.Y.R. some years earlier, and Stanier's advent later on the L.M.S. is a story well told. But there were other ways of making a change, and Fowler and Anderson imposed the standards of the erstwhile Midland Railway upon the whole new L.M.S. group in 1925, not as a result of any comparative assessment or due to their especial merits for future development, but as a last fling of the besetting sin of the old individual railways which was, in an age of minimum mutual contacts, to believe devoutly that only one's own practice could possibly be the best. Few of these standards survived the Stanier regime, and hardly any found their way on to the British Railways standard series.

Churchward followed by Collett on the G.W.R. and Stanier on the L.M.S. achieved in their incumbency as C.M.E., completely integrated series of locomotive designs covering the whole traffic range. On the other two post-grouping railways, the L.N.E. and the Southern Railways, only a partial standardisation took place. Gresley as is well known, retained and continued to have built a number of classes to the designs of former constituent companies, differing down to the smallest details from his own highly standardised family of designs for the more important duties. On the Southern, under Maunsell, the larger engines were developments of Urie designs for the former L.S.W.R., while the smaller were all based upon the Ashford school of the S.E.C. Railway. Each of the two groups was highly standardised as to boilers and details within itself—practically nothing was standard as between one group and the other.

It would be difficult to prove that this divergence from the pure doctrine produced any less effective motive power for its respective railways than did the more academic standardisation of the other two major systems, and this draws attention to the point that the advantages of total standardisation in the locomotive world, although real, are not overwhelming. Really spectacular advantages only arise when mass production becomes possible, mass production in the sense of 200,000 identical units in a year, as in the automobile industry. Total annual programmes consisting of even the most highly standardised locomotive classes have rarely exceeded 400 engines on British Railways and it has never been demonstrated in clear and unequivocable figures the extent to which there is a measureable difference in annual capital and maintenance charges as between full and partial standardisation, so long as in the latter case there is a sufficient run of any given class over several years, and a sufficient total stock to justify its own reasonable provision of spares.

One reason why the difference does not emerge clearly arises from the inherently long life of steam locomotives. It is usually 10 years or more before a standard series infiltrates the locomotive stock of a given railway by as much as 50 per cent, and in the meantime stocks of different kinds of parts have to be stored to maintain the remaining non-standard engines.

The meticulous accounting which would be required to identify the relative maintenance costs and allocate them correctly to standard and non-standard stock, adequately weighted against the productivity of individual locomotives, has so far not been thought worth while having regard to the cost that would be involved.

The position can, therefore, be summarised by saying that in the locomotive field, standardisation has obvious advantages on the basis of engineering instinct and judgement, and that it has been accepted in whole or in part by the majority of locomotive engineers from the middle of last century onwards. On the other hand, its value has never been rigidly proved, and its potential economic gains tend to get lost in a fog of overlapping previous and subsequent standards, the more so to the extent that any one of these has continued into obsolescence.

Outside the main stream of unification as outlined above, some very strange things have been standardised on railway engines. Such was the wheel spacing of 8 ft. and 8 ft. 6 in. between centres on six-coupled engines first laid down by Johnson in the last century for the Midland Railway, which became the obligatory norm for all further Derby design, stretching into the 1930s. Whether appropriate to the type of locomotive or not, and irrespective of the fact that the associated coupling rod designs might vary due to differing power requirements and cylinder dispositions, the draughtsman was instructed to lay down first on his clean sheet of drawing paper, wheel centre lines at this sacred spacing and to carry on from there. All 0–6–0, 0–6–2, 2–6–2 and 2–6–4 engines schemed or designed over a 50-year period were thus distinguished.

Another curious thing to standardise across a number of types of differing capacities is the grate area. Having achieved a boiler of harmonious proportions and satisfactory grate area of 26 sq. ft. on his 4–4–0 engines, Robinson of the Great Central Railway, or whoever it was who thought out these things on his behalf, adopted this as standard for his subsequent 4–6–0 engines of four different classes, in none of which did it prove to be sufficient against the intended duties. Not only so, but in each case the ashpan arrangements were such as to impede seriously the flow of air through the firebed as ash built up with the result that steaming was fickle, and coal consumption enormous.

A particularly disastrous standardisation was the insistence by Anderson of the widest possible use of the axlebox from the Derby Class 4, 0–6–0 freight engine—itself designed pre-World War I—on to types built for the L.M.S. after grouping. Two important new classes, the Garratt and the Class 7, 0–8–0, each justifying a modern and generous axlebox design in its own right, were completely spoiled by the affliction of this particular idea, and they both had to be prematurely scrapped due to the miserable annual mileage they attained between repairs due to their being inadequately shod.

It is with diffidence that one refers in this context to Gresley's decision to standardise the three-cylinder principle for locomotive propulsion. On his larger locomotives he made a notable success of this and produced performances which have never been surpassed in this country. But while all kinds of reasons were advanced for extending the application in doctrinaire fashion to his smaller engines, no particle of factual proof has come to light that they were any the better for it compared to what they would have been as straightforward two-cylinder machines, cheaper in first cost and maintenance. An objective examination of "Bantam Cock" the three-cylinder 2–6–2 which, had Gresley lived, was intended to become the widely extended medium mixed traffic class of the future, can only lead to the conclusion that the two-cylinder 4–6–0 B1 class engine which Thompson, his successor, ultimately produced for this purpose was a much sounder conception.

Reverting to that part of the definition of "standard" which refers to locomotives specifically identified for use as a series on more than one railway system, it is true that in the early years of railways, private manufacturers often produced their own standard designs which they sold to a number of companies in identical form other than the painting and lettering. This custom died away, however, with the emergence of the railways' own design policies and building works, and it has only been resurrected in quite recent years, where three out of the five tyes of British Railways standard diesel locomotives are to the design of private manufacturers specifically intended for general use in their respective power classifications throughout the whole of British Railways. A variant of this emerged in 1872, when Crewe Works supplied ten 2–4–0s and 86 0–6–0 tender engines and five four-coupled shunting tanks to the Lancashire & Yorkshire Railway, all exactly to Ramsbottom's designs for the L.N.W.R., except for copper-capped chimneys and green paint. In the words of Ahrons,* "There is no knowing how long Crewe Works would have continued to oblige by manufacturing for another railway, but the private locomotive building firms were unkind enough to obtain an injunction on the grounds that a railway company had powers only as a carrying and not as a manufacturing trading concern."

In later times mention can be made of a standard series of locomotives for British Railways as a whole which was considered at the end of World War I, when it was in the balance as to whether nationalisation or large-scale amalgamation would fall to the lot of the individual railways which had served the country up to that time. There was also another and different standard series for the lines comprising the newly formed L.M.S.R. mapped out in 1924, while Hughes was still C.M.E., and before midlandisation became an established fact. These "might have beens", of only academic interest now, would greatly have changed the flow of British locomotive development had they come to pass as visualised at the time.

When I was working in Horwich drawing office in the latter part of the first war, we were aware that George Hughes was concerned with A.R.L.E. activities directed towards evolving proposals for locomotives for general use, over all railways, but no diagrams of which I am aware were con-

* "Locomotive and Train Working in the latter part of the Nineteenth Century", L.Y.R., E. L. Ahrons. (*Rly. Mag.*, Vol. XLI, 1917.)

tributed from the L.Y.R., a probable reason being that, since the loading gauge of that railway was the most generous of any as regards height, nothing based upon its practice was likely to be acceptable dimensionally. Eventually there came into the office the first results of the work done by Clayton at Ashford for Maunsell on behalf of the A.R.L.E. in the form of two fairly detailed diagrams for a 2–6–0 and a 2–8–0 respectively. These were handsome and well proportioned proposals clearly based upon the S.E.C.R. N class 2–6–0 which had come out in 1917, but with one or two significant differences. Parallel instead of taper boilers were used while the boiler pressure was kept down to 180 lb./sq. in.—evidence that George Hughes had taken some part in the discussion. But the outcome would have been very inimical to route availability, since it resulted in the cylinder diameter of 19 in. on the N class being increased to 20½ in. on the proposed standard 2–6–0 and to no less than 21½ in. on the 2–8–0, dimensions which in conjunction with their horizontal setting would have precluded use of these engines over most of the lines forming the eventual L.M.S. and all of the eastern lines of the Southern. Those of us interested in locomotive development were all agog to see the passenger version of these schemes, but the possibility of nationalisation faded away at that time and with it all idea of a common motive power stud. The diagrams were lost to view for many years until, as a result of a question during the discussion on C. S. Cocks' paper on the history of Southern Railway locomotives in 1948, they were resurrected and published on pages 859 and 860 of the Inst. of Locomotive Engineers Journal No. 206. A later reference to them with a further reproduction of the diagrams was included in Vol. 1 of Holcroft's *Locomotive Adventure* (p. 88).

Some other curiosities in diagram form illustrating phases in this A.R.L.E. development were also published in an article by E. B. Trotter in *Trains Illustrated* for February 1954. This included such curiosities as a very G.W.-looking 2–6–0 with round-topped, domed boiler and outside Walschaert gear, stated to have been prepared at Swindon in August 1917, and four other proposals for 2–8–2T, 2–8–0, 4–4–0 and 4–6–0 engines ascribed to Derby in 1918, all with improbable stove pipe chimneys.

Although I was posted to the Derby drawing office eight years later, and most assiduously combed through all the old diagrams then remaining of abortive projects, some of them dating long before 1918, I never came upon any trace of these particular "A.R.L.E. standard" proposals. This by no means proves, of course, that they never existed.

The other series of standard proposals referred to was outlined in 1924 for general use on the newly formed L.M.S. Railway and is set out on Horwich Drawing No. 17039 dated 23/4/24 reproduced in Fig. 1*. As I made this drawing under instruction, and as all documentary evidence concerning it has long since vanished, a few comments from memory may be of interest. In my former book I have referred to the design work which George Hughes initiated at the end of 1923 with a view to providing locomotives for the newly formed group, and in particular described the

* Reproduced as front end-papers.

development of proposals for a 2–6–0, a 4–6–2 and a Garratt for heavy freight work. By early 1924 preliminary outlines for such engines had been committed to paper, and the 2–6–0 scheme had got as far as incorporating a Belpaire boiler such as was already in use on the L.N.W. "Prince of Wales" class, but modified to increase the grate area and in certain other details. At this time Hughes wished to explain to the Euston management how this initial work could be expanded to cover all traffic requirements for the whole railway, and the 12 types set out on the above mentioned diagram formed a first rough cockshy for preliminary discussion. The presentation is of interest as showing how little the route availability problem was realised in those early days, and unlike the eventual British Railways 12 standard types shortly to be described which spanned six distinct ranges of axle weight and route availability, Hughes' effort, excluding the "Pacific", was clearly intended to meet only three such levels. Another feature was a survival of the one time almost universally held belief that engines needed tailoring to particular routes and duties by fine graduations in wheel diameter and adhesive weight, leading to passenger engines having both 6 ft. 9 in. and 6 ft. 3 in. wheels four- and six-coupled, and freight engines having 5 ft. 6 in. and 5 ft. 1 in. wheels. The three variants of main line tank engines are a case in point, and the future was still veiled when it would be realised that with modern cylinder, valve gear and bearing design, a single type could do duty for all.

These engines were obviously intended to be highly standardised as to boilers, cylinders and all details, and, again excepting the "Pacific", were to be straightforward two-cylinder machines following the lead of the 2–6–0 development, whose boiler and cylinders were destined intact for several of the types only with the stroke increased from 26 in. to 28 in. A little more detailed outlining was undertaken in respect of two of them and Figs. 2 and 3 reproduce individual diagrams which were prepared for the 4–6–0 and the 2–6–4 tank. In the process the wheels of the former were increased from 6 ft. 3 in. to 6 ft. 6 in. Even so, bearing in mind the successful performance of the 2–6–0 when it came out in 1926 and the fact that it combined long travel valve gear with high superheat, it is possible to regret that the L.M.S. was not given the advantage of having these classes built, and it was not until 1934 that the "Black Staniers" provided a 4–6–0 machine of the same kind of all purpose utility.

It was clear, however, that Euston management was under fire from the Follows—Anderson—Fowler axis which had an entirely different engine building philosophy to peddle, and which won the day in the end. Thus a potentially interesting standard series proved no more than a flash in the pan and quickly perished.

There now remain in this preliminary review of earlier standard series, only what may loosely be termed the involuntary standards—namely locomotives built in quantity for wartime purposes, and which either during or after the war were run over a large variety of lines in Great Britain. The earliest of these was the Great Central two-cylinder 2–8–0 locomotive widely multiplied for overseas use in the 1914-1918 war. As a simple and

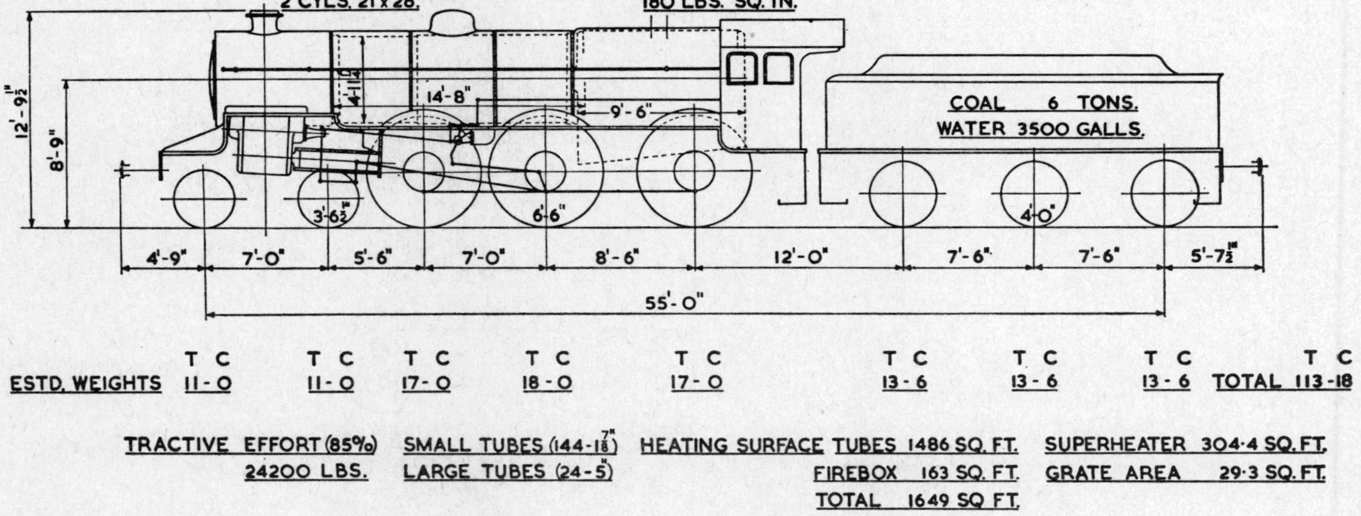

SCHEME FOR 4-6-0 TO RUN ON MIDLAND DIVISION & IN SCOTLAND USING 2-6-0 CYLINDERS AND MODIFIED PRINCE OF WALES BELPAIRE BOILER.

Fig. 2.

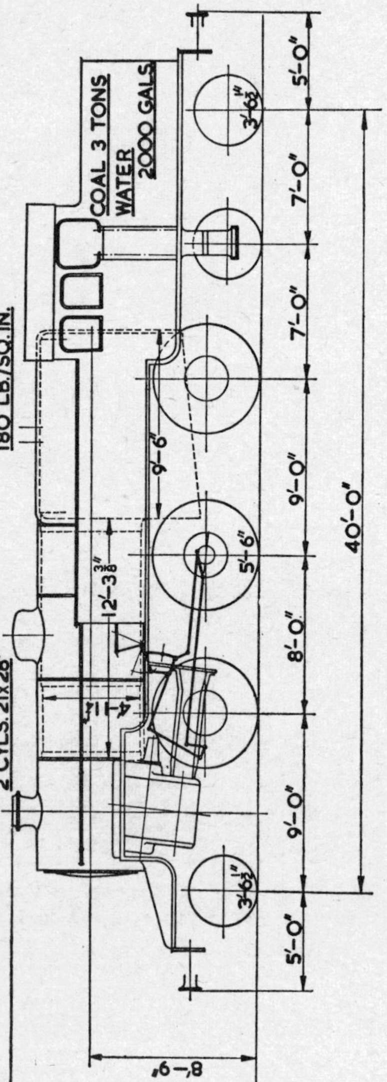

	T C	T C	T C	T C	T C	T C	T C
ESTD. WEIGHTS	11-0	19-0	19 0	18-0	11-10	11-10	TOTAL 90-0

TRACTIVE EFFORT (85%) 29280 LB.

SMALL TUBES (144-1⅞") LARGE TUBES (24-5")

HEATING SURFACE TUBES 1308 SQ. FT. FIREBOX 163 SQ. FT. TOTAL 1471 SQ. FT.

SUPERHEATER 304·4 SQ. FT. GRATE AREA 29·3 SQ. FT.

SUGGESTED SCHEME FOR 2-6-4 TANK ENGINE TO USE SAME PARTS AS 2-6-0 TENDER ENGINE (HORWICH DIAGRAM No. 17020)

Fig. 3.

very robust machine it was eminently suitable for this new purpose, but when it came back home, or worse still as the mounting momentum of new production was caught short by the end of the war, literally hundreds of these engines were to be seen idle and forlorn in sidings about the country. They were too wide across the cylinders to run over the majority of railways as they then were, and although a bold distribution was made of twenties and thirties to individual railways, those forming the eventual L.M.S. found it impossible to use them due to dimensional restrictions. I remember how interested we were at Horwich to watch the arrival of such alien creatures, and it was on one of these that I made the first footplate trip of my life, on a foray into Yorkshire, in which it was intended to see by watching the deformation of lead "fingers" attached to the cylinder covers, how near we came to fouling platforms—alas it was too near. Eventually as is well known the newly formed L.N.E. Railway absorbed a large number, and the G.W. Railway acquired 100; with its generous loading gauge it was the only other company who could run them usefully.

As a tribute to their rugged construction no less than 92 of these old stalwarts were transferred to the War Department in 1941, and were sent overseas for service in the Middle East. None of them were returned to this country after the war.

In 1939, the Ministry of Supply selected the standard Stanier 2–8–0 of the L.M.S. as being the most suitable for use with the forces overseas, and placed orders on outside firms for 240 such engines of which however only 208 were actually built as under:—

	Ordered	*Built*	
North British Company	100	158	1940-1942
Beyer Peacock	100	50	1940-1941
Vulcan	40	—	
	240	208	

The cancellation of the outstanding 32 engines and the rearrangement of the orders between builders presumably resulted from the switch over to production of "Austerity" locomotives shortly to be described.

In addition to these 208 engines, 51 more built at Crewe or by Vulcan for the L.M.S. in 1936/7 were also transferred to the War Department. Of this total of 259, 31 never went abroad and were allocated back to the L.M.S., 23 were lost at sea and 161 remained overseas, leaving 44 which were returned to the U.K. after the war, some as late as 1957. As a result of this whole transaction 51 engines had been given up by the L.M.S., but it had received back 75, a net gain of 24.

Entirely separate from the above was the result of a further Ministry decision in 1942 that the type of heavy freight locomotive constructed by all British Railways for domestic working should conform in future to this type—it being noted that it was the only existing design of this kind which was capable of unrestricted running over the majority of lines of the four main line companies. Accordingly it was arranged that on completion of existing orders for their own types, the L.N.E., G.W. and Southern railways would commence building the Stanier engine. In the outcome 313 were so built in the years 1942-1946, none of which went overseas and all of which had by 1947 homed to their parent the L.M.S. This construction was divided as follows:—

	Built by	*Built for*		
		L.M.S.	*L.N.E.*	*Total*
L.N.E.	Doncaster	30	20	50
	Darlington	30	23	53
G.W.	Swindon	80	—	80
S.R.	Eastleigh	23	—	23
	Ashford	14	—	14
	Brighton	68	25	93
	Total	245	68	313

In all 852 of this "standard" engine were built from 1935 onwards by railway and private workshops for the railways themselves or for the War Department (Plate 2). The saga of their wanderings, their allocation and reallocation is most complex, but when it was all over two were retained by the W.D. at Longmoor for training purposes, 161 remained abroad either destroyed or serving Middle East Railways, 23 were lost at sea in transit and 666 were taken into L.M.S. stock. This account is not altogether irrevelant to our subject, for although not ultimately working on any other railway than the L.M.S., at any rate until recently, the fact of their having been built by every major locomotive building works in the country had its effect upon policy decisions regarding the eventual B.R. standard locomotives as will presently be seen.

While the above measures sufficed for the conditions of the earlier part of the war where action was confined to far flung but limited fronts in the Near East, the prospect of the second front and the invasion of Europe opened up an entirely new scale of probable requirements, and this coincided with a serious shortage of labour available for locomotive building, when the large-scale diversion to the forces and to Government war work was considered.

R. A. Riddles, one time Principal Assistant to the C.M.E. of the L.M.S., had become Deputy Director General, Royal Engineer equipment, and amongst his other duties, he was responsible for the procurement of railway material for the army. Posed with the conflicting circumstances outlined above, he proceeded to develop a simplified version of the L.M.S. 2–8–0 in which effort he was assisted by the loan of T. F. Coleman, Chief Locomotive Draughtsman of the L.M.S. in a consulting capacity, and by the design resources of certain of the private locomotive manufacturers. The Stanier 2–8–0 called for 33,000 man hours in its construction and included in its 92 tons of empty weight was 17 tons of steel castings and 16 tons of forgings, both in competition with urgent armament requirements. The principal changes were:—

(1) Round top parallel boiler in place of Belpaire taper boiler.

(2) Replacement of all steel castings and a few of the forgings by fabricated plate structures or by cast iron.
(3) Cast iron in place of steel wheel centres.
(4) Trimming feed from oil boxes in place of mechanical lubrication for axleboxes.
(5) Displacement in place of mechanical lubrication for cylinders.
(6) The two-bar crosshead was replaced by the "Laird" type having twin bars arranged above the piston rod. This was to reduce wear due to lodgement of sand on the bottom bar which had adversely affected Stanier engines in the Near East.
(7) Cylinders modified to accommodate six above plus certain other simplifications to aid production.
(8) Large eight-wheel tender.
(9) Simplified cab and footplate arrangement.

As it was Riddles who was at the helm, there had to be also an unexpected twist which took the form of a squat little chimney, lower in height than the other boiler mountings, thus proclaiming the work of an individualist!

Combined with these design and materials changes, was the contribution of the workshops production experts of the private manufacturers who between them evolved simplified methods of manufacture even for unchanged components, so that in total a saving of 6,000 man hours, roughly 20 per cent was obtained over the original locomotive.

Stanier was a bit restive about this transformation of his brain child, considering that there might have been too great a sacrifice of features which contributed to good operation and economical maintenance. As a good engineer fully aware of all the circumstances, including the short potential life of W.D. engines on the battlefield, he accepted the position however. But when the Ministry issued some publicity about the new engines containing a good deal of ballyhoo, Stanier was offended by reading into it a veiled implication that what had been jettisoned from his own design was worthless, and that it had taken smart fellows at the Ministry to see this and act accordingly. This had not been, of course, stated in actual words, and may not even have been intended, but the impression was there, and it did not please him at all. I have by me as I write, a long analysis I had to do for him at the time, setting out all the pros and cons for making an official protest, but luckily nothing came of it.

In 1944 a 2–10–0 version of the W.D. 2–8–0 was produced with the object of getting down from a 15½ to a 13½-ton axle load to meet special service conditions, and at the same time to favour combustion of low grades of coal by use of a wide firebox providing 40 sq. ft. of grate area in place of the previous 28.6 sq. ft. The tractive effort and all other components of the two classes remained identical.

A third wartime product of the Ministry was a 0–6–0 shunting saddle tank, in this case based upon a standard design which the Hunslet Engine Co. of Leeds had produced for colliery and other industrial service. Of very simple straightforward design, this engine had 18 in. × 26 in. cylinders, 4 ft. 3 in.-diameter wheels, a working pressure of 170 lb./sq. in.

Table I
Building of W.D. Austerity Locomotives 1943-1946

Type	*Makers*	*1943*	*1944*	*1945*	*1946*	*Total*
2-10-0	North British Loco Co.	15	85	50	—	150
2-8-0	North British Loco Co.	243	249	53	—	545
	Vulcan Foundry	160	180	50	—	390
					Total	935
0-6-0T	Hunslet Engine Co.	44	57	19	—	120
	Andrew Barclay	—	—	9	6	15
	Robert Stephenson	30	39	21	—	90
	Vulcan Foundry	—	—	50	—	50
	Hudswell Clarke	14	21	12	3	50
	W. G. Bagnall	—	29	10	13	52
					Total	377

and a tractive effort of 23,870 lbs. It weighed 48 tons in working order.

Nine hundred and thirty-five, 150 and 377 of these three "Austerity" types were built by private manufacturers as shown in Table 1 making a

Table II
Dimensions of Early 'Standard' Freight Engines

	Stanier 2-8-0	*W.D. 2-8-0*	*W.D. 2-10-0*	*Ashford 2-8-0*	*Hughes 2-8-0†*
Cylinders Dia. x Stroke	18½″x28″	19″x28″	19″x28″	21½″x28″	21″x26″
Coupled Wheels Dia.	4′ 8½″	4′ 8½″	4′ 8½″	4′ 8″	5′ 1″
Tractive Effort lbs. (at 85% B.P.)	32,438	34,215	34,215	37,500	30,650
Weight: Engine only in working order in Tons	72·2	72·0	78·6	75·0	—
Boiler Working Pressure lbs/sq. in.	225	225	225	180	180
Outside Dia. Front	5′ 0″	4′ 7⅛″	4′ 7⅛″	5′ 6″*	5′ 3¾″
Back	5′ 8⅜″	—	—	—	—
Between Tubeplates	12′ 2⅞″	12′ 0″	15′ 8″	14′ 0″*	12′ 2″
Large Tubes No. x Dia.	21 5⅛″	28 5⅛″	28 5⅛″	—	24 5⅛″
Small Tubes No. x Dia.	202 1¾″	193 1¾″	152 1⅞″	—	161 1⅞″
Heating Surface					
Tubes sq. ft.	1,479	1,512	1,759	1,644	1,345
Firebox	171	168	192	158	160
Total	1,650	1,680	1,951	1,802	1,505
Superheater	230	310	423	268	307
Grate Area sq. ft.	28·6	28·6	40	30	27·5
Valves Diameter	10″	10″	10″	11″	11″
Steam Lap	1½″	1½″	1½″	1½″	1½″

*Approximate—scaled from diagram.
†Intended to use same boiler and cylinders as on Harwich 2-6-0 which eventually became an L.M.S. standard type.

total production of 1,462 engines*. Seven hundred and thirty-three of the 2–8–0s, 25 of the 2–10–0s and 75 of the shunting engines were acquired by British Railways after the war, and were used widely as truly standard engines capable of running almost anywhere.

* A very complete account of the inception, building and operation of the "Austerity" tanks was given in *The Locomotive* for October 14th, 1950.

Plates 3, 4 and 5 illustrate the engines, and Table II compares their principal dimensions with those of the proposed Ashford 2–8–0 standard design of 1919 and of Hughes projected standard of 1924.

Although we are now dealing only with the outposts of the central locomotive standardisation which we are shortly to describe, two matters of some interest arise at this point. One is to know what was in fact sacrificed in performance and efficiency by all the simplifications which transformed the Stanier into the W.D. 2–8–0s, both subsequently adopted as widely used standard types. Secondly, as the cylinders and valve gear and wheels of the W.D. 2–8–0 and 2–10–0 engines remained the same, what was the relative effect of the very different boilers which were fitted. A series of constant speed tests carried out in 1953 by the Derby testing unit with the L.M. mobile test plant, enables these questions to be answered. Table III gives the results of the tests carried out on the W.D.

Table III

Comparative Performance and Efficiency as Disclosed by Constant Speed Tests

Class of Engine	WD.2–10–0	WD.2–10–0	WD.2–8–0	WD.2–8–0	Cl.5 4–6–0
Coal					
Colliery	Blidworth	Blackwell	Blidworth	Blackwell	Firbeck‡
Grade	2B Hard	3B Soft	2B Hard	3B Soft	2B Hard
Calorific Value					
As received BTU's/lb.	12,800	11,800	12,800	11,800	13,000
Max. Drawbar HP on level	1,380	1,250	1,020	650	1,315
Min. Coal per DBHP/hr.lbs.	2·5	2·79	2·69	2·87	2·30
Min. Water per DBHP/hr.lbs.	18·5	18·5	19·1	20·0	16·5
Max. Steam Temperature °F.	650	635	610	*	680
Evaporation, Actual at Front					
end limit lbs/steam/hr.	27,000	23,500	19,500	13,000†	23,050
Corresponding lbs/coal/hr.	4,360	4,000	3,240	1,870	3,440
Min. Coal per 1HP/hr. lbs.	2·04	2.30	2.12	*	1·80
Min. Water per 1HP/hr. lbs.	15·3	15·3	15·4	*	13·9
Efficiency %					
Boiler at					
1000 lbs./coal/hr.	81	81	84	*	83
2000 lbs./coal/hr.	76	76	75	*	76
3000 lbs./coal/hr.	70	70	65	*	69
4000 lbs./coal/hr.	64	65	—	*	—
Maximum Output	63	65	63	*	66
Cylinder Thermal	12·6	12·6	12·4	*	13·6

†Could only be sustained for 45 minutes.
*Not recorded.
‡This coal has the same characteristics as Blidworth.

engines. Unfortunately, the Stanier 2–8–0 was never subjected to similar constant speed tests, but the corresponding Class 5 4–6–0 was most thoroughly tested. The cylinders and valve events were identical, the firebox and boiler diameter were also identical between the two engines, but the Class 5 had a boiler barrel 1 ft. longer, and instead of a 21-element superheater, had 28 elements, which in fact brought it into line in this respect with the W.D. engines which also had 28 elements. The last column in Table III although quoting results from the 4–6–0 engine, can be taken to represent within a very small margin of error, what the corresponding Stanier 2–8–0 was capable of achieving.

Reverting to the first question postulated above, it will be seen that the simplifications on the 2–8–0 have exacted a price in performance and efficiency, starting with the steam-producing capacity of the boiler and draughting arrangements inferior by some 16 per cent, which when coupled to a reduction in cylinder thermal efficiency, resulted in a lowering of the maximum horsepower available at the drawbar on the level by $22\frac{1}{2}$ per cent. The 28 elements on the W.D. engine only produced a maximum steam temperature of 610°F. against 680°F., and fuel and water consumptions whether on an indicated or a drawbar basis were up by from 11 to 18 per cent. Allowing any desired margin for the fact that the 4–6–0 was not a 2–8–0, the point being made here is not to criticise the W.D. engine, whose performance and efficiency was ample for the kind of wartime duties it was specifically designed for, but to indicate that the detailed improvements initiated by Stanier and further developed by Ivatt, were in fact very worth while.

As to the second question, the great superiority of the wide firebox boiler on the 2–10–0 is plain to see from Table II, whether as an economical steam-producing instrument or as a means of digesting a poor quality coal very prone to the formation of hard clinker.

This digression has turned aside to inspect one or two of the interesting side issues which have arisen incidentally from the design of locomotives for common purposes, but to revert to the main stream of thought in this chapter, and to sum up what is to be deduced from all that had been done in this country with standard locomotives before 1948, enough has been said to make three points as under:—

(1) Standard locomotives are not a new idea but a very old one indeed however the word standard is interpreted.

(2) Most locomotive engineers have sincerely believed that there is something to be gained from standardisation. If a specific proof of this is lacking there is equally no clear evidence that it is not so.

(3) Table II begins to indicate what has become progressively the strongest argument for standardisation by types, namely that steam locomotive development had in recent years reached a stage in this country where the more external design varied, the more the fundamentals remained essentially similar.

To amplify the last point above, who will say that any benefit could arise from one over another of the differing external design of the four 2–8–0s set out in Table II spaced through the years from 1919 to 1942. Once they were tailored to meet the required route availability conditions, then any one could have proved as well as any other capable of dealing with heavy freight all over the country, and could have proved just as receptive of internal design improvements in draughting arrangements, boiler ratios, steam, gas and air circuits, and all the other ingredients leading to the ultimate of what the straightforward Stephensonian steam locomotive is capable.

This background was very much in the minds of those who were placed in charge of locomotive engineering when at the end of 1947 the four remaining separate railways in Great Britain were combined into a single entity under nationalisation.

II

Setting the Stage—1948

THE TRANSPORT ACT which received the Royal Assent in August 1947, was a formidable document of 128 sections and 15 schedules, but its intention was unmistakable. It created the first British Transport Commission upon which it laid the duty of providing "an efficient, adequate, economical and properly integrated system of public inland transport and port facilities for passengers and goods". The Act further provided that there should be public authorities known as Executives, and, by agreement with the Minister, five of these bodies were established, of which one was the Railway Executive. In the words of Sir Cyril (later Lord) Hurcomb, first Commission Chairman, "the Railway Executive had to begin by welding four very large railways, each with its own techniques and traditions, into a single working unit. . . . But substantial changes in organisation, policy and practice are inevitable if the Commission are to seize their opportunities for greater economy and higher efficiency which unification brings, and which it is their duty to seize. To ensure that these changes be thought out and brought about in reasonable time, the Railway Executive has been appointed in part on a functional basis . . . the emphasis is functional because it is desired to secure, first, the earliest possible analysis and comparison of differing company techniques, and procedures, with a view to standardising the best. . . ."

There was the remit in unequivocal terms "to compare existing techniques and standardise the best". It is as well to remind ourselves how very clearly this requirement was stated at the time.

To carry out this policy in motive power matters, R. A. Riddles was appointed member of the Railway Executive for mechanical and electrical engineering, and in the closing months of 1947 he set about selecting the team of principal officers who were to join him at Marylebone on January 1st, 1948, the first day in the existence of British Railways. In a pioneering position of such potential power, the background and character of the first incumbent was clearly bound to have the utmost importance in what followed, and would inescapably mould the direction of future events. Riddles had begun his engineering training at Crewe Works in 1909 on the old L.N.W.R. and had occupied a number of positions subsequently in the workshops at Crewe and Derby until 1933, and then became Locomotive Assistant and two years later Principal Assistant to W. A. Stanier, Chief Mechanical Engineer of the L.M.S. Railway. In these latter positions he was closely concerned with developing the locomotive standardisation policy which Stanier had initiated in 1932. In 1937 Riddles became Mechanical & Electrical Engineer for Scotland, and on the outbreak of the second world war, joined the Ministry of Supply where he was called upon to create a directorate for the provision of all Royal Engineer equipment, including that for transport. He became Deputy Director General of this organisation in 1941, and during his service with the Ministry he initiated and took responsibility for the design and construction of the 2–8–0, 2–10–0 and 0–6–0T W.D. locomotives for immediate use in the different war theatres, and for ultimate use on British Railways. In 1943 he returned to the L.M.S. Railway as Chief Stores Superintendent, and became in 1946 a Vice-President.

The highlights of his career were thus an extensive knowledge of workshops, and all that a locomotive should consist of to facilitate economical workshop procedures, secondly first-hand experience in the conception, design and construction of standard locomotives, and thirdly an unusually wide range of administrative practice.

The functional direction which had been ordained from above, meant that in future rolling stock design, and the means by which, and the standards to which, it should be maintained should now be taken away from the four Chief Mechanical Engineers and two Chief Electrical Engineers who had previously directed such matters on the former railway companies, and these activities henceforth became the responsibility of the new department set up at 222 Marylebone Road. Riddles in his Presidential Address to the Institution of Locomotive Engineers on November 16th, 1950,* has explained in detail what this department consisted of and what were its aims. Of standardisation he had this to say, "Technical standardisation has been widely sought on British railways since the 1860s and each successive amalgamation has widened the field of application. Moreover in railway experience it cannot be said that standardisation has prevented progress, because the standards have in general been kept fluid and adaptable. On the technical side one of the virtues of nationalisation was to remove all remaining barriers. The bulk of engineering opinion agrees that progressive and enlightened standardisation is a good thing, for it does reduce first costs, maintenance costs and stores stocks. There have been, and are, bold spirits who will have none of it, but they are in a considerable minority. If then standardisation is good, the removal of barriers to it is good, whether this be done under nationalisation or private enterprise. There undoubtedly were barriers under the four former companies."

Under Riddles, two administrative posts were set up which directly affected locomotive policy, these being a Chief Officer (Locomotive Construction and Maintenance), and an Executive Officer (Design). For the first of these he had no hesitation in nominating R. C. Bond, with whom he had worked closely over many years on the L.M.S., and who, although trained at Derby, had the closest affinity with his experience and outlook. Because of this selection, and certain others from the same

* *Proc. Inst. Loco. E.* No. 218, 1950.

stable which had been made for positions outside locomotives themselves, Riddles was under some pressure to choose his design man from another railway. Candidates from the Western and Southern were strongly canvassed, but his mind was set on recommending the author, who again had been closely associated with him and with Bond in much of the design and development work undertaken on the L.M.S. from 1934 onwards. With such an identity of background, experience and outlook these three men were likely to work in harmony and to get results, and above all to support one another through thick and thin in all the complex activities which now lay before them.

Under the chief, design was in future to come very much within the province of the Executive Officer (Design) whose duties were assigned by Riddles in the words "He co-ordinates design requirements for locomotives, carriages and wagons laterally with the Technical, Operating and Commercial Chief Officers (at H.Q.). He also controls a two-way traffic of allocation of design vertically downwards into the regional offices, and of co-ordination of experience from the regional offices back to headquarters, so that decisions at that level can be broadly based. He is also the connecting link between the locomotive design offices on the one hand and the carriage and wagon design offices on the other, which otherwise by the nature of their work follow a rather independent course. This liaison assures common standards wherever these are applicable."

The foregoing leads to the conclusion that the introduction of a series of standard steam locomotives for the newly formed British Railways was inevitable. Given nationalisation itself, the policy which was laid down by the Transport Commission and its chairman, the nature of the duties laid upon the Railway Executive, the nature and experience of the man who became its member for mechanical and electrical engineering, and finally the henchmen whom he chose to carry out his intentions, given all these things, then no other outcome was even remotely possible.

There has, of course, been much after criticism to the effect that, granted standardisation was in the air, the wrong things were standardised—steam locomotives instead of diesel or electric. There is a great deal which can be said on this subject, most of which has already been ventilated freely and loudly, and it would swell this book to unreasonable proportions to repeat it all here. Let us concede that the time was hardly ripe. The establishment of British Railways coincided with a Government White Paper on capital investment which deferred the prospects of large-scale electrification. Moreover much spade work required to be done before even a unified system of current supply for the whole country could be agreed. Of diesels, there were at take-over date, only 49 shunters and one main-line locomotive in stock, L.M.S. No. 10,000. The shunters were standardised, but no one in his right mind was prepared to base any policy of main-line diesel traction at that time upon so slender a foundation as a single locomotive which had only been running four weeks before the end of 1947.

It was thus fixed in the minds of the triumvirate that there should be a standard series of new locomotives for the new railway and that they should be steam. But what kind of steam locomotives? Three steps were initiated almost at once to find out. The first was a series of Interchange Trials which would record the leading features of performance and efficiency of the principal existing locomotives over the main lines of all regions, the second was to set up a locomotive standards committee which inter alia was to examine the possibility of selecting a limited range of existing standard types of the former companies for construction in 1950 and onwards until such time as new standard types were developed. The third measure was to initiate work on the nature of the proposed new standard locomotives themselves. All three tasks were completed by September 1948, and it is proposed now to deal in some detail with each in turn.

It was an obvious pre-requisite to considering any standard locomotives to establish the capabilities and relative efficiencies of existing types, and above all to ascertain whether or no variations in design were called for to meet different geographical features as distinct from different categories of traffic, passenger, freight, or mixed, heavy or light.

In the first four months after the setting up of the Railway Executive, there was accordingly organised a series of tests, in which five passenger, four mixed traffic and five heavy freight engines (14 types in all) were run in revenue service to the best current schedules and with maximum loads over the routes selected. Table IV summarises the pattern which was evolved. Since potential universal service was at issue, a single quality of coal—best Midlands hards—was used throughout, but the locomotives were each operated by footplate crews from the owning regions, assisted

Table IV
Pattern of Interchange Trials 1918

Routes	*Locomotives*				
	E.R.	*L.M.R.*	*W.R.*	*S.R.*	*W.D.*
E.R. King's Cross–Leeds	A4	Duchess Royal Scot	King	Merchant Navy	—
Marylebone–Manchester	B1	Class 5	Hall	West Country	—
Ferme Park–Peterboro'	O1	Class 8F	28XX	—	2-10-0 2-8-0
L.M.R. Euston–Carlisle	A4	Duchess Royal Scot	—	Merchant Navy	—
St. Pancras–Manchester	B1	Class 5	—	West Country	—
Toton–Brent	O1	Class 8F	—	—	2-10-0 2-8-0
W.R. Paddington–Plymouth	A4	Duchess Royal Scot	King	Merchant Navy	—
Bristol–Plymouth	B1	Class 5	Hall	West Country	—
Acton–Severn Tunnel Jc.	O1	Class 8F	28XX	—	2-10-0 2-8-0
S.R. Waterloo–Exeter	A4	Duchess Royal Scot	—	Merchant Navy	—
Eastleigh–Bristol	O1	Class 8F	28XX	—	2-10-0 2-8-0
Sc.R. Perth–Inverness	B1	Class 5	—	West Country	—

when on foreign metals by "conductors" from the home railway who knew the road.

To record performance and obtain some measure of efficiency, a dynamometer car was attached to each trial train, and as only three were available, the work was split up amongst them so that they and their crews were employed to the best advantage. It was realised that this would not achieve absolute comparative efficiency for which stationary test plants or a system of controlled road tests in steps of constant steam rate would be necessary. Neither were however as yet available, and even if they had been, the time required would have been too long. Under general direction from headquarters the day-to-day organisation of the trials, the working out of the results, and the preparation of the final report, were vested in an ad hoc committee consisting of Messrs. B. Spencer, C. S. Cocks, S. O. Ell and R. G. Jarvis, who at that time were most closely connected with locomotive performance and testing on the Eastern & North Eastern, Southern, Western and London Midland Regions respectively.

The trials took place between April and September 1948, and as can well be imagined, they raised a great deal of professional and public interest at the time, and they were widely reported. C. J. Allen later published two volumes* which described objectively and in the closest detail all that took place.

Eighteen years have passed since then and for those who were not interested in these trials at the time, or to whom the above mentioned books are not available, it may briefly be recalled that the best all-round running performances were achieved by the L.M.S. "Rebuilt Scots" and the Bulleid Pacifics. The latter were, however, much inferior in efficiency to any of the others, and this was not due to the enterprising manner in which they were driven, but to inbuilt design characteristics. The lowest consumptions in lb. per DBHP/hour were attained by the L.N.E. A4 Pacifics, which however almost alone amongst the contestants, suffered mechanical breakdown on three occasions, due to hot inside big ends. The Stanier Pacific came second as regards low specific coal consumption, but was driven throughout in a very unenterprising manner, so that its running performance was far below the potential which other records have proved this class capable of attaining.

Of the mixed traffic engines the L.N.E. B1 performed better than anticipated and the L.M. Class 5 rather worse, but both displayed a similar and satisfactory efficiency. Owing to the different nature of their duties and the operating vagaries inherent in goods train running, the freight engines showed a wider scatter of results and there was little to choose between them other than a tendency for the W.D. 2–8–0 to come out bottom in any relative assessment.

The Western engines were rather in a class by themselves. Their physical dimensions limited their use to the L.N.E. and Western lines so

* *The Locomotive Exchanges*, C. J. Allen (Ian Allan, 1949); *New Light on the Locomotive Exchanges*, C. J. Allen (Ian Allan, 1950).

that they did not run the full course. Their performances can be described as adequate rather than brilliant, certainly not up to that noted by various recorders under other circumstances, and their efficiency was naturally lower than that attained by their brethren having 100°F. or more increased superheat. Subsequent separate trials on their home ground sought to show that they had also been penalised by having to use Yorkshire coal. These later trials indicated that the extent of this debit was as much as 12 per cent, but subsequent work on the stationary test plants long afterwards, by isolating and quantifying the many different aspects of combustion, have indicated that the intrinsic difference in consumption between these two coals, other things being equal, need not be more than that of the relative calorific values namely 6 per cent.

There is only a little more which can usefully be said about the interchange trials from the point of view of those who initiated them and in the context of the ultimate British Railways Standard locomotives.

The objectives of the Railway Executive were necessarily not quite the same as those of the railway enthusiasts who took so much interest in the trials. As previously explained, the former was bent upon exploring whether there was any reason why one type of locomotive was not as effective as any other on appropriate duties in alternative geographical situations, whereas the latter viewed the trials as a sporting event in which they did not wish to see their favourites handicapped in the slightest degree.

Criticisms were largely centred round the latter aspect, and it was said that some engines were penalised in performance because drivers were not precisely instructed as regards the making up of time. Those who lost time on the not too difficult schedules and who made little effort to regain it, were thought to have won an unfair advantage in coal consumption over others who blasted away and produced brilliant performances. The enterprise or caution of the conductors were claimed to have had an influence on the handling of their engines by drivers who had only had a week's preparatory running over each route before the trial began. It was pointed out that the official train weights were at tare and did not include the weight of passengers and luggage which varied from day to day and might even equate to an additional coach. Moreover, the performance of the operating departments on certain lines, notably the L.M. west coast route, was dreadful; slow trains were routed in front of the test trains with insufficient headway and little effort was made to ensure clear runs. It did not escape notice that the Western Region engines were made to eat Yorkshire coal which it was claimed they did not like, instead of the Welsh coal on which they were nurtured.

All of these things were true enough, and there were reasons for each which it would be tedious at this late date to explore. The majority evened themselves out amongst the variety of drivers and driving practices, routes and operating methods, so that it is difficult to feel that injustice was done to any particular contestant, or that the course of history would have become much deflected had more attention been paid to some or all of them.

1 Great Central design of 2–8–0 as built for the War Department during the 1914-18 War. [L. King

2 Stanier 2–8–0 No. 48490 on an up freight train approaching Ampthill Tunnel on the Midland main line south of Bedford. [E. S. Cox

3 Austerity 2–8–0 No. 90042 at Stratford, Eastern Region. *[R. E. Vincent*

4 Austerity 2–10–0 No. 73788, now WD 601 *Kitchener,* at Polmadie, Glasgow, in July 1957 when on loan from the Longmoor Military Railway. *[A. W. Martin*

5 Austerity 0–6–0 saddle tank No. 68051 in original B.R. livery.

7 Class 7 4–6–2 No. 70004 *William Shakespeare* with a special exhibition finish for the "Festival of Britain" in 1951.

6 L.M. design of 2–6–4T built at Brighton for use on the Southern Region.

8 Class 6 4–6–2 No. 72000 *Clan Buchanan.*

9 Class 5 4–6–0 No. 73072, with flush sided tender.

10 Class 9 2–10–0 No. 92087

11 Class 8 4–6–2 No. 71000 *Duke of Gloucester*.

12

Class 4 4–6–0 No. 75066.

13

Class 3 2–6–2T No. 82037 in B.R. passenger livery.

14

Class 3 2–6–0 No. 77009.

15 Class 4 2–6–0 No. 76000.

16 Class 4 2–6–4T No. 80010.

17 Class 2 2–6–0 No. 78000.

18 Class 2 2–6–2T No. 84002.

19 A Class 7 boiler on the riveting tower in Crewe Works.

Inner copper firebox of a Class 7 locomotive.

Ashpan of a Class 7 locomotive.

22 Cut-away perspective view of a Class 7 engine showing the firebox, grate and ashpan arrangement

24 An earlier proposal for a Class 7 locomotive with a double chimney and bar frames.

23

Self-cleaning plates and spark arrestor netting in the smoke-box of a Class 7 engine.

25

Cab end view of No. 70000 showing the injectors and other fittings attached to the boiler.

26 Walschaerts valve gear on Class 7 4–6–2 No. 70000 *Britannia.*

27 Reversing gear showing the screw acting directly on the weigh shaft.

28 Completed main frames for a Pacific locomotive.

The benefits and shortcomings of these tests from the point of view of Riddles and his men lay in somewhat other directions. They did confirm what had already been gathered from experience on the L.M.S. that a properly designed steam locomotive would work satisfactorily anywhere from Lands End to John o' Groats on appropriate duties. They showed that the simplest and most straightforward designs were at no disadvantage compared with the more complex. They indicated that capacity to boil water was as essential as it ever was, and that with modern cylinders and valve events, choice of number of cylinders, other than as demanded by loading gauge, was of little importance, and that relatively small wheels were no bar to high speed.

With the benefit of wisdom after the event, it can now be seen however, that we were misled in certain directions by the essentially artificial characteristics of all such trials. For example, employment of the most experienced drivers of whom the regions were possessed, masked the true facts of adhesion. We concluded that uncompensated 4–6–2 engines could be almost as sure-footed as 4–6–0s not realising that it was the supreme expertise of individual drivers who coaxed seemingly reliable adhesion out of the various Pacifics even when on mountainous routes. In the hands of "run of the mill" drivers and with indifferent weight adjustment on individual axles, such engines could become most uncertain starters. The end coupled locomotives on the other hand sat down at the back end the more firmly on the track, the greater the drawbar pull, and the Pacifics could only match their adhesive reliability in the hands of all comers when their axle weights were compensated throughout as was the case abroad. The sure-footedness of the Western "Kings" during the trials was thus a true indication, but that of the various Pacifics was not fully representative of what they could be like under day-to-day conditions.

Taking all things together in this series of interchange trials, rough justice was done in the time available, and the results enabled us to proceed with confidence in what we proposed to do. Both for our own developments and for the history of the British steam locomotive, these trials contributed a valuable record.

In parallel with the above, and initiated by a minute of the newly constituted Mechanical Engineers Committee at its first meeting on January 8th, 1948, a Locomotive Standards Committee was established with the following remit:—

"(1) *Design of Consumable Details*

(a) To prepare a list of renewable details which it is desirable to standardise from the point of view of purchase and manufacture.

(b) To consider existing designs of such details and recommend the best for adoption on future new construction.

(c) To make recommendations as to how far, if at all, it is practicable to apply such standards to existing motive power units.

(2) *Recommendations for* 1950 *Construction*

The possibility is to be examined of selecting only one of the existing standard types in each of the operating categories listed below, for construction in 1950 and onwards until such time as new standard types are developed.

In order that the selection can be made on rational grounds, the Committee should assemble as much information as is available under the following headings in respect of each of the existing standard types:—

Power capacity of engine and boiler.
Mileage between consecutive classified repairs.
Casualty record.
Record of tests already carried out.
Costs so far as they are obtainable on a uniform basis.

There will also be available test results and experience arising from the interchange of locomotives shortly to be carried out.

The Committee will require to obtain particulars of dimensional and weight limitations, so that the engine types selected may be able to run over the largest possible number of routes, and they should also consider what special fittings and details, if any, are desirable on the selected types in order to ensure maximum availability and efficiency.

The objective is to have a considered recommendation available in each of the traffic categories indicated below, before June 30th, 1948, so that the building programme can be formulated and material ordered before the end of this year for 1950 construction. The amount of information which is collected in support of the recommendations should be conditioned by the need to adhere to this date."

The traffic categories referred to above were:—

4–6–0 or 4–6–2 Mixed Traffic Tender (as L.N.E. B1 class).
2–6–0 or 4–6–0 Mixed Traffic Tender (as L.M.S. Class 4).
2–6–2 or 2–6–4 Heavy Mixed Traffic Tank (as L.N.E. L1 class).
2–6–2 Light Mixed Traffic Tank (as G.W. 45XX class).
2–6–0 or 0–6–0 Light Freight Tender (as L.M.S. 6400 class).
0–4–0 Dock Shunting.

There were also some riders to the remit covering similar studies of diesel and electric traction on the mechanical side, but these were not followed up.

To undertake this very comprehensive enquiry Riddles was sensible enough not to nominate any of his own staff who might be subsequently subject to a charge of bias. Instead he appointed A. W. J. Dymond of the Swindon office to be Chairman, with C. S. Cocks representing the Southern Region, J. W. Caldwell, the L.M. and E. Windle the E. & N.E. Regions, on the main committee. To deal with the standardisation of details, two sub-committees were formed, one for boiler mountings, and the other for miscellaneous items. Here again a Western Region man was appointed to the chair in each case, with a member of the drawing office staff from each of the other regions making up the remainder. It was something of a master stroke to place responsibility for this phase of the work in Western Region hands, for they had had less experience of previous large-scale amalgamation, and consequential inter-railway standardisation, and they would thus be introduced to the facts of life in this context in a very practical manner.

Now there are some points about the task appointed to these men worth commenting upon, before the outcome is recorded. The examination and selection of standard details was more or less straightforward, but item 2 was to cover an eventuality which might or might not arise. Nineteen fifty was the earliest date after nationalisation at which any change could economically be introduced into what had already been authorised by the Boards of the former railways. It was not then clear how long it would take to evolve and design the series of B.R. standard locomotives on which Riddles mind was set. If the time was short then continued building of the current designs of each region would fill the gap. If for any reason several years were to elapse, then the possibility of concentrating building upon only one out of the four regional designs in each traffic category was certainly worth exploring.

The list of traffic categories contained in the remit may look curious, but the crying need following the war years was to strengthen the moderately-sized mixed traffic fleet throughout the system, of which some 40 per cent had become over age. The individual railways had already between 1945 and 1948, substantially strengthened their express passenger fleets, while the tremendous influx of heavy freight engines built during the war, both company and War Department, eliminated any immediate needs in this sector.

Riddles knew as well as anyone that it was unlikely that the full remit as regards repair mileages, casualty records and costs could be achieved, not only due to differing quantities of information available, but also due to the widely differing bases upon which available figures were founded in the different regions. He hoped, however, that this would prove a means of uncovering anything there might be of value in this connection within existing records. He also knew that only some existing L.M.S. designs and a few from the Southern were suitable for universal use in Great Britain, due to axle weights and spacing, and width over cylinders, and cab roofs as affecting Civil Engineers route availability. However, he wanted this aspect to be thrown up impartially in the report. Last of all, he hoped that this difficult remit with its strict time limit would test the mettle of the men nominated to deal with it, and thus identify promising men for future promotion.

In place of the mountain of comparative information which it was hoped to elicit, only a molehill of a report was forthcoming on May 31st, 1948. It examined 22 existing designs closest in character to the types nominated in the remit. It recorded for each, diagrams, dimensions, hammer blows, tractive effort curves and equivalent uniformly distributed bridge loadings, but had nothing to say on repair mileage, casualty records, records of tests already carried out or on costs. The part of the Committee's work concerned with standardisation of details was incomplete at this date, but formed the subject of later and very useful reports. On the central question of selection from amongst existing types for interim building and on the possible application to these of standard details and fittings, it had this to say:—

"The Committee have reviewed all the evidence brought forward in respect of the locomotives referred to above, and have to report that although the possibility exists of selecting one in each of the six traffic categories for building in all regions, they do not feel justified in making such a selection for the following main reasons:—

(1) None of the engines submitted exhibit characteristics which exclude any type on the grounds of inferiority in design and performance.

(2) The fundamental requirements of route availability and power characteristics for each type of locomotive for use in all regions have not been defined.

The Committee are, moreover, of the opinion that the proposal to build, as an interim policy until future British standard designs are evolved, types of engines for use in all regions which are already in use on one or other of the regions, while attractive at first sight, is nevertheless not desirable from a long-term point of view. In reaching this opinion the Committee are influenced by the following considerations:—

(i) In each region main workshops and motive power depots are equipped with tools, facilities and stores suitable for the building, repair and maintenance of the regional stock of locomotives, and in very large measure standardisation within the regions has been achieved.

(ii) The introduction of new British standard designs of locomotives, incorporating new standard components, will necessarily result in appreciable increase in tools and stores in main workshops and motive power depots, due to the requirements of the new locomotives.

(iii) Any major change in design standards inevitably involves some degree of inconvenience and delay in repair and maintenance with subsequent adverse effect on locomotive availability.

(iv) Were selected existing designs to be adopted in all regions for 1950 and subsequent building programmes until the new British Railways standard designs were built, two changes of practice and procedure in main workshops and motive power depots in most if not all of the regions would be entailed. The effect of two such changes would have an enhanced adverse effect on repair and maintenance.

(v) Some, if not all of the regions would have to build, and afterwards maintain and operate for many years, engines which conformed to neither the existing regional standards nor to the new British standards.

For these reasons the Committee recommend:—

(a) That the proposal to build for general use in 1950 and subsequently until new British designs are evolved, one of the designs in each traffic category included in the remit be abandoned.

(b) That until British standard designs for use in all regions are evolved, each region continues to build engines to its existing designs."

In fairness to the Committee members, it must be conceded that little other result than the above was likely to be forthcoming. Each man of them had been for long years the servant of C.M.E.s whose only wish was to be left alone to pursue their former ways, and this strong loyalty naturally overpowered any incipient "imperial" thinking which their appointment to this early committee of the newly-formed Railway Executive was able to encourage. On the surface the conclusions and their arguments were entirely logical, but it did not escape the notice of Riddles or his associates, that most of the arguments against selection of individual existing types were also arguments against the development of the new standard locomotives themselves. What the committee failed to note was that few if any of their arguments had carried any weight on their own railways in the past when C.M.E.s wished to improve the breed by either partial or total alteration. Moreover their workshops and motive power depots were able to achieve remarkably good availability figures and economical repair with no less than 440 different existing designs already, legacy from past changes. The addition of up to a dozen further types either selected from other regions or new standard types was a drop in the ocean by comparison.

In the outcome, building of local types for the individual regions continued until 1954, and a total of 1,550 non-B.R. standard locomotives were built subsequently to January 1st, 1948. It is interesting to note moreover, that notwithstanding the committee's solemn warnings, several L.M. Region types were in fact selected for building on other than the home region, and were allocated to regions which had never used them before. Thus on the Southern for example, a line specially suited for the use of tank engines, the breaking up of life expired indigenous types created needs which no other existing types on the region were available to fill, and 41 L.M.R. Fairburn Class 4, 2–6–4 tanks built in S.R. workshops (Plate 6) together with 30 Ivatt Class 2, 2–6–2 tanks built at Crewe, were added to Southern Region stock in the years 1950-1952. Similarly on the Eastern & North Eastern Regions there was a crying post-war need for modern small six-coupled freight engines to replace numerous old 0–6–0 types. While there existed at the end of 1947 a proposal in diagram form at Doncaster for a small 2–6–0, a kind of scaled-down K1 class, there seemed no need to design and produce this new class when Ivatt's Class 4 and Class 2, 2–6–0s could satisfy the traffic need in all respects. The motive power departments concerned arranged transfers and trials from the parent region, and 100 of the two classes were built at Darlington and Doncaster between 1950 and 1952, and a further 10 built at Horwich, all of which were allocated to these two regions. Finally 25 of Ivatt's Class 2, 2–6–0s were built at Swindon in 1952/53 for use on the Western Region. Table V summarises the ultimate position.

Table V

Construction and Original Allocation of L.M.R. Locomotives to Eastern, North Eastern, Western and Southern Regions, 1948-1952

Built At	*Original Allocation by Regions*				
	Eastern	*North Eastern*	*Western*	*Southern*	*Total*
Crewe				41290–41319	30*
Horwich		43122–43131			10*
Darlington	43080–43095 43104–43106	43070–43079 43096–43103			37
	46465–46469	46470–46482			18
Doncaster	43058–43069 43107–43111 43142–43161	43050–43057			45
Swindon			46503–46527		25
Brighton				42066–42106	41
Total Number	61	49	25	71	206

*Crewe and Horwich built many others for L.M. and Sc. Regions.

Note – 43XXX are Class 4, 2–6–0
46XXX are Class 2, 2–6–0
41XXX are Class 2, 2–6–2T
42XXX are Class 4, 2–6–4T

All of this was accomplished with much acceptance to the operators, and with none of the woe to workshops and depots which the committee foresaw.

So much then for the second of the three exploratory measures which Riddles introduced. The last, which consisted of a first examination of what a new British Railways standard series might comprise, is so closely bound up with the eventual development of the locomotives which are the main subject of this book, that it is proposed to refer to this in the next chapter.

III

The Twelve B.R. Standard Types—their development and realisation

THIS CHAPTER first of all discusses the reasons why these locomotives took the form they did, describing the background to their design and the main features which it was desired to incorporate in their make-up. It goes on to refer to the different factors which varied the original intention during the course of design, and, type by type, it arrives at the form of these locomotives as they were finally constructed and put into service.

In the first place the human background is relevant. When Riddles during a visit to Derby towards the end of 1947 let me know that I had been chosen as one of his team, and that I was to be put in charge of design, he told me that my main job would be to undertake design work on a new series of standard locomotives, carriages and wagons. Being wise enough not to install a dog, and then do the barking himself, he did not specify just what form this standard rolling stock was to take. He did not need to, for we both knew that on the locomotive side, the prime objective was to continue the work successfully developed by Ivatt and his team in the latter days of the L.M.S. towards simple robust machines which, while incorporating the results of available technical development, were nevertheless primarily economical units to build and maintain in sheds and workshops.

Bond, who was henceforth to direct maintenance standards and practice, also knew from past experience which ingredients were most likely to attain these objectives and ease his future responsibilities, and here again no great discussion or detailed specification was necessary to the designer, for the long Bond/Cox association had established close identity of views. Thus with a minimum of direction I started as soon as I arrived at Marylebone to assemble a first report on the proposed standard steam engines as a basis for discussion with all interested parties.

Now some words as to the foundations upon which this exercise was based. Few men are totally independent of the influences which nurtured them, and it was natural therefore that I was influenced by L.M.S. thought and practice. Not blindly so however, for as my previous writing may have shown, I was as strongly critical of L.M.S. practices when things were going badly, as any more independent observer was likely to be. On the other hand it did seem to me, and this has not been seriously contradicted so far as I know, that in its last few years the L.M.S. had definitely stepped ahead of other companies in those matters which were to be the keystone of our new endeavours. To the groundwork laid in Stanier's time in raising boiler repair performance to altogether higher levels, in exorcising the blight of hot boxes, broken tyres, and lateral instability in running, was added, under Ivatt, the manganese axlebox liner with its strong effect on repair mileages, and the application of the self-cleaning smokebox, rocking grate, and self-emptying ashpan, which transformed manual requirements at the sheds and permitted to steam a longer lease of life against a background of increasing labour shortage for disagreeable tasks. In addition the lessons taught by Wagner of the former Reichsbahn on boiler proportions, and by Chapelon of the S.N.C.F. on opening out the steam circuit had been fruitfully studied. All this besides the more obvious retention of high superheat, long valve travel valve gear of robust proportions, and piston valve design which minimised internal leakage, added up to a body of practice which had brought the steam locomotive to its highest level of economical operation and maintenance so far. By simple inspection, the three other former main-line companies had adopted this or that feature out of the above list, but none of them were using all.

None the less engineering interest, if not common sense, demanded detailed knowledge of what design consisted of on other railways at home and abroad. As regards the other British companies there had never been difficulty in obtaining any desired information ever since the close association of the C.M.E.s and their staffs when serving the Railway Executive Committee during the last war. The measures described in the last chapter were designed to explore to the utmost what there existed of value on these railways, both as regards complete locomotive designs and also the details of which they were composed. To supplement this I made it my business not only to visit the regional Drawing Offices, and to attend personally as many of the interchange trials as other duties permitted, but also to undertake wide footplate riding on all the principal types of the former companies. Apart from the Western's retention of low superheat the plain fact emerged that, however much the locomotives differed superficially and in details, there was little to choose fundamentally between the recent products of Swindon, Doncaster or Derby, outside the greater development at the latter centre of certain features referred to above. They followed the same principles of combustion and draughting, worked over broadly the same steam temperature range, delivered their power at the rail by similar mechanical means, and were each as subject as the others to the human factor in driving and firing, and to the integrity of examination and maintenance. So also were Southern Region locomotives in the main but a new factor had arisen with the advent of Bulleid on that line, who had introduced some novel features into his locomotive designs since 1941. We naturally had to give these especially careful consideration, for if any of them did in fact represent a breakthrough from traditional and conventional practice, then we should be remiss in not applying them to our own designs. If performance alone were the only criterion, then, as the interchange trials emphasised, Bulleid's engines were at the top of the form. But enough experience had been gained by

1948 to demonstrate without question that this was obtained at too high a price, and that from the point of view of reliability, ease of servicing, periodicity of repairs and cost of maintenance to say nothing of specific fuel consumption, his "Pacific" locomotives were far inferior to the corresponding products of other designers. Since we were committed to develop the best all-rounders, and not only lively performers, no case could be made for adopting the more controversial components from his designs, such as chain driven miniature valve gears, or oil baths, but as will be seen when details come to be described, there were nevertheless some other items from Bulleid's designs which we were glad to make use of.

Apart from these general considerations, there were more personal influences. So compelling is the "mystique" of the steam locomotive that all of its devotees worthy of the name have their own private ideas as to what it should consist of and look like—obviously influenced by what they have seen and experienced while in thrall to their obsession. As early as 1919 I was cursing cylinders and valve gear inside the frames, as I struggled to assemble these parts under L. & Y. 0–8–0s on hot summer days. By 1941 I had learned that there were more solid objections to inside cylinders than mere discomfort, as I studied the whole subject of the dynamic forces arising from different cylinder dispositions.* As a youthful railway enthusiast, it was always the engines with the outside cylinders and valve gear and the raised footplating which took my fancy—the Maunsell and Gresley 2–6–0s, the Urie 4–6–0s on the L.S.W., and almost all American locomotives. Then in later years and especially during mechanical inspection in sheds and shops, the troubles with inside big ends on three-cylinder engines had been powerfully impressed on my mind. Visits to Germany in 1937 and to America in 1945, reinforced preferences both as to simplicity and as to appearance, while the study I was asked to make in 1942 on the three-cylinder engines of the L.N.E.R., and even the mechanical behaviour of the A4 Pacific on the interchange trials all combined to intensify a yearning for the least number of cylinders to do the job.

What I have written above of personal preferences dovetailed completely into the ideas of Riddles and Bond, who were equally active during 1948 in making the rounds of British Railways as a whole and in formulating workshop and maintenance policy. The same kind of locomotive which appealed to me as designer, appealed to them from quite independent considerations of policy, construction, and maintenance.

Locomotive design however is far from being a party game, and however enthralling these preliminaries were to the triumvirate, they were not sufficient in isolation to define the new standards, for various external bodies were naturally and powerfully concerned in what we were doing. For example the Railway Executive had immediately set up an "Ideal" stock committee, following L.M.S. precedent, composed mainly of

* "Balancing of Locomotive Reciprocating Parts", E. S. Cox, *Proc. Inst. Loco. E. Journal* No. 165, 1941.

operating and financial officers, whose duty was to recommend the volume and value of rolling stock replacement year by year and to establish a norm of ultimate stock against which annual replacement programmes were to be directed. This committee dealt with global numbers in power categories and affected us by defining the power classes which were needed and the order and pace at which they were to be provided. In this formative period also Workshops, Motive Power and Operating Officers volunteered or were called upon to submit statements of the features they wished to see incorporated in future builds. These were carefully examined, but as can well be imagined were sometimes contradictory and cancelled one another out. The unions also, the N.U.R. and the A.S.L.E. & F. tendered views as to what their footplate members would like to see, but here again we could establish little concrete from such submissions, as they tended to be somewhat parochial; a fact not to be wondered at when the close identification of drivers and firemen with their regional motive power is considered. It was very necessary that the above parties, as users of the locomotives to be, should give their opinions, but it was rather disappointing how little that we could definitely use emerged from their consideration.

Another department vitally concerned in what we were doing was that of the Civil Engineer. Most fortunately the quarrels and acerbities of the past had long since abated, and from the first there was most excellent co-operation and harmony between the two sides at any rate so far as rolling stock route availability was concerned. In particular the loading gauge problem was tackled with speed and precision, and by the time production design was required to begin we had available what were known as the L_1 and L_2 locomotive loading gauges. The former offered almost unlimited running over the main lines of the newly formed regions, whilst the L_2 gauge permitted some increase in cross section for the most powerful locomotives at the cost of limiting their use to the more important main lines. Apart from this, permissible axle weights and wheel spacing were identified, and the myriad levels of bridge and permanent way conditions handed down through time from the former companies were rationalised into steps of axle loading related to route availability. When our design work commenced it was necessary to consider axle load categories of 22½, 21, 19, 17, 16 and 13½ tons according to the different lines over which the new engines would be required to work. To these six divisions of axle weights were added the results of a meeting of minds on the hammer blow question, in that the mechanical engineer was prepared to reduce his percentage of reciprocating weights balanced to moderate values, while the civil engineer conceded a generous interpretation of the relation between hammer blow and axle load.

This outline of the specification for design process seems a far cry from the days when an autocratic C.M.E. designed and had built what he thought fit, and the rest of the railway, management, operators, maintenance men and motive power sections, had to like it or lump it. But such success as attended this way of doing was balanced by some horrible failures as history attests, and at the level of costs, and the far from monopolistic conditions of the post-war period, more democratic methods were insisted upon, and indeed were no more than the application of good horse sense to engineering. Before leaving these preliminaries, it may be that the lay railway enthusiast will wonder where in all of this discussion is the item of performance. After all the prime purpose of any locomotive is to move passengers and freight at competitive speeds, and the question may well be asked. "If all this was defined; economical construction and maintenance requirements, standardisation, the needs of operators and staff, and satisfaction to the civil engineer, where does the general public come in, whose satisfaction in speed and punctuality makes the passenger part of the business possible?" The simple answer is that so far had the art of locomotive design progressed by 1948, that almost any locomotive, if it followed the broad rules which had become so well established, could not help but perform adequately, consistent with the capacity designed into it, and on the assumption that it was properly driven and fired and was well maintained.

Against this background we now come to the Report on Proposed Standard Steam Locomotives which I completed on June 14th, 1948, a report which was thus summarised in its introduction:—

"It proposes 12 types to cover the major traffic requirements of British Railways, which fall into three main groups as under:—

(1) *Entirely New Designs*

*Class 6 4–6–2 Express Passenger
„ 6 4–6–2 Mixed Traffic
„ 5 4–6–2 Mixed Traffic
„ 8 2–8–2 Freight

(2) *New Designs which are developments of Existing Types*

Class 7 4–6–2 Express Passenger
„ 4 4–6–0 Mixed Traffic
„ 3 2–6–2T Mixed Traffic

(3) *Existing Designs modified in Detail*

Class 4 2–6–0 Mixed Traffic
„ 4 2–6–4T Mixed Traffic
„ 2 2–6–0 Mixed Traffic
„ 2 2–6–2T Mixed Traffic

The above types make use in varying combinations of standard boilers, flanging blocks, cylinder patterns and wheel and tyre sizes. It is proposed that they shall all use the same range of standard fittings and details.

* The class numbers here referred to are those of the former L.M.S. Railway. In the early days of nationalisation, in applying this system to the whole stock of B.R. engines, class numbers from 6 upwards were altered. The former 5X class became the new 6, and former classes 6 7 and 8 became 7, 8 and 9 respectively. Class 5 and below remained as before.

Three main considerations lie behind the proposals here put forward:—

(a) That entirely new designs should be put forward wherever a definite step forward in availability and efficiency can be realised.

(b) That new designs should not be undertaken for their own sake, and that where an existing design offers all that is available from the present state of design, it should be continued with only detail modifications.

(c) That the whole trend should be towards simplification, good accessibility of all parts requiring attention, and reduction in time required for repairs and servicing.

It is important that all the designs should be readily adaptable to the incorporation in future years of the results of work undertaken at Rugby Testing Station, and by the other testing apparatus which is, or will become, available to British Railways. Some of the simpler of the resulting improvements will be applicable retrospectively to all the Standard engines, while others can be incorporated in successive mark numbers of the original basic design.

Consideration has been given in formulating these proposals, to the views of the Operating Superintendents of all regions, and to the suggestions for detail improvements and amenities which are being sought by the Unions. Indications from the current locomotive interchange trials have also been considered so far as they are presently available."

The first group of four entirely new designs was proposed as a means of assuring to the "second eleven" of railway motive power, as it were, a definite step forward in availability, ease of servicing, mileages between repairs, and efficiency.

By "second eleven" was meant that large group of six-coupled locomotives in power classes 5, 6 and 7 and of 2–8–0s in Freight Class 8. All existing regional examples, except only the S.R. West Country, had narrow fireboxes, while the more powerful of them had multi-cylinders. It was proposed that they should be superseded by new designs all of which would carry wide fireboxes, first of all to raise maximum steam-producing capacity against possible increase in future speed of working with given loads, and secondly to promote lower rates of combustion for given power output with good coal, and to ease the problem of obtaining good steaming with poor coal. It was further proposed to ease building, servicing and maintenance problems by eliminating inside cylinders altogether, since the desired tractive efforts in these power classes were obtainable within the new L_2 loading gauge by two-cylinders only.

Accordingly Fig. 1 in the Report, reproduced herein as Fig. 4 outlined four locomotives, of which the first two were identical, except for a difference of 6 in. in wheel diameter, and which, having a maximum axle load of 21 tons, were visualised as suiting Class 7 (old Class 6) duties for passenger and mixed traffic services respectively. The remaining two engines which were to carry the same boiler, smaller than the above, were respectively a Class 5 4–6–2 mixed traffic type with 18 tons axle load, and a 2–8–2 freight engine with $16\frac{1}{2}$ tons on the coupled axles. Apart from the two-cylinder arrangement and the wide fireboxes, bar frames were common to all, as was a standard outside bearing carrying wheel used on bogie, trailing truck and tender. The four locomotives were proposed to replace for future building the "Castle", "County", "Hall" and 28XX classes on the W.R., the "Rebuilt Scot", Class 5X and Class 5 4–6–0s and 2–8–0 classes on the L.M., B_1 4–6–0 and o_1 2–8–0 on the E.N.E. and the "West Country" and Q classes on the Southern.

These proposals were a first shot at interpreting the remit which we considered to have been laid upon us, and in the many discussions which followed with all concerned, a number of circumstances arose which led to considerable modifications. These can be listed as under:—

(1) It was quickly perceived from the interchange trials and other trials carried out regionally at about this time that there was going to be no justification for a step of as little as 6-in. wheel diameter as between two otherwise similar types. Not only had the Bulleid "Pacifics" demonstrated how, with a modern front end, a 6 ft. 2 in. wheel was no bar to speeds up to 90 m.p.h., the civil engineers permitted maximum, but some tests on the E.N.E.R. between A_1 and A_2 Peppercorn Pacifics 6 ft. 8 in. and 6 ft. 2 in. wheels respectively, also indicated in a higher power category, what little need there was for so fine a distinction. It was accordingly resolved to telescope what had been originally designated types 70 and 71 into a single Class 7 design having 6 ft. 2 in. coupled wheels for all duties within a limited axle load of 21 tons, and having a boiler with 42 sq. ft. of grate area.

(2) The proposed smaller "Pacific" in the 18-ton axle load Class 5 range became the subject of particular heart searching because it was a border line case. On the one hand, truly Class 5 duties were already carried out with great competence by the numerous 4–6–0 mixed traffic locomotives already existing, having total weights some 11 tons less than that of the proposed 4–6–2. At this point Bond took an active hand in the deliberations, pointing out that the lower combustion rates which would result from the larger grate at the average rate of working on such duties would give savings almost exactly equated by the capital charges on the increased cost of the Pacific over the 4–6–0, whilst the higher standby losses and the maintenance cost of the trailing truck would remain as debits. On the other hand, if the main idea behind the new engine was some increase in potential capacity beyond that of the existing 4–6–0s, as indeed it was, then it would be more appropriate to develop it as a Class 6 from the first. The performance of the S.R. West Country of this power category in the recent trials had been very impressive, and although its design was not acceptable due to its poor efficiency and mechanical complexities, it was clear that there was scope for an $18\frac{1}{2}$-ton axleload locomotive of this kind.

It was thus decided to proceed with two designs, the Class 6 having the same chassis and wheel spacings as the Class 7 but with a smaller boiler

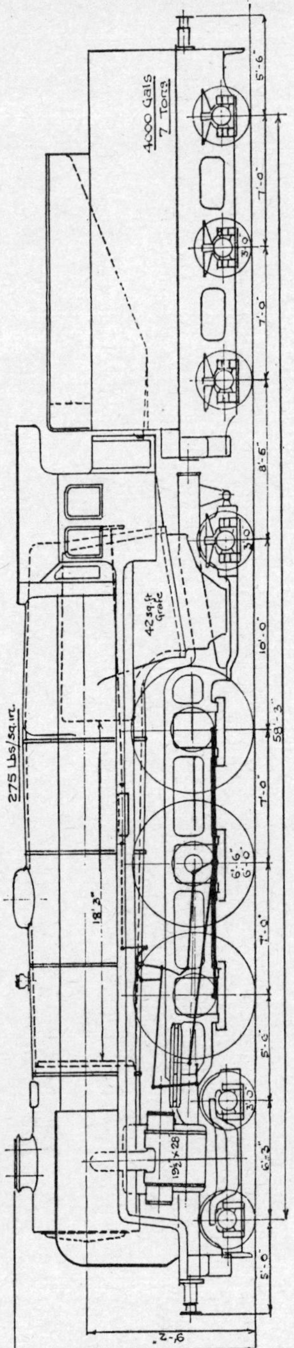
4000 Gals
7 Tons
42 sq ft Grate
275 Lbs/sq.in.
18'3"
19½"×28"
58'-3"
5'-6"
7'-0"
7'-0"
8'-6"
10'-0"
7'-0"
7'-0"
5'-0"
6'-3"
5'-0"
3'0"
6'-6"
6'-0"
9'-2"
13'-2"

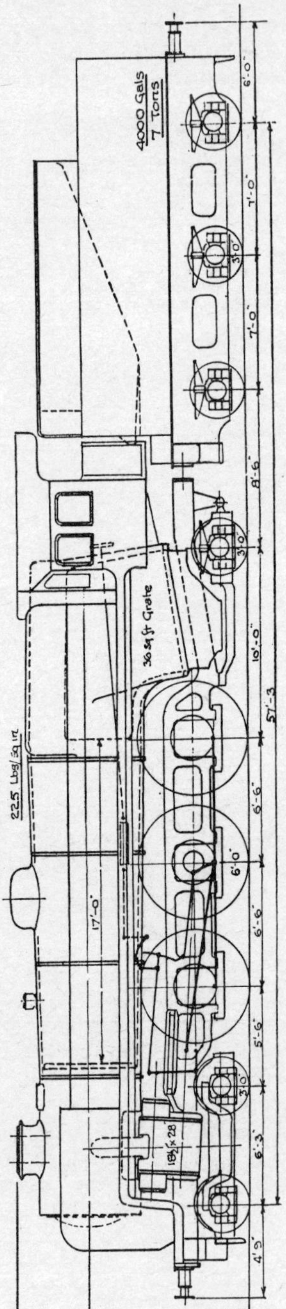
4000 Gals
7 Tons
36 sq ft Grate
225 Lbs/sq.in.
17'-0"
18½"×28"
57'-3
6'-0"
7'-0"
7'-0"
8'-6"
10'-0"
6'-6"
6'-6"
5'-6"
6'-3"
4'5"
3'0"
6'0"
8'-9"
12'9"

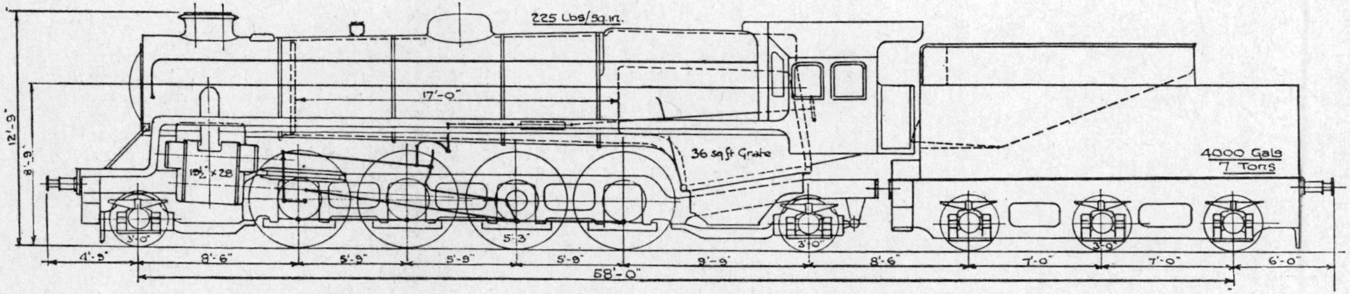

Fig. 4. Original proposals for 4–6–2 and 2–8–2 locomotives. From top: Type 70 (passenger) and and Type 71 (mixed traffic) class 6 with 21 ton axle-loading; Type 72 (mixed traffic) class 5 with 18 ton axle-loading; and Type 90 (freight) class 8 with 16½ ton axle-loading.

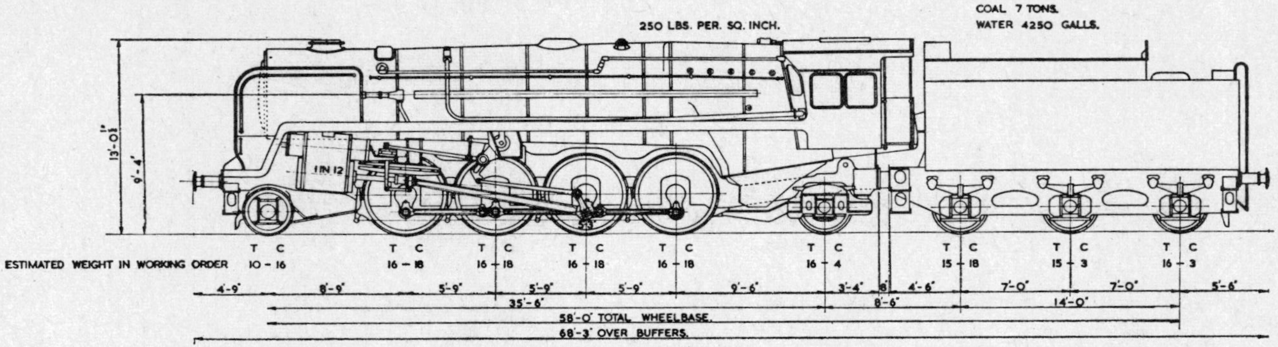

BOILER BARREL DIAMETER (OUTSIDE)	5'-9" INCREASING TO 6'-5½"
FIREBOX (OUTSIDE)	7'-0" LONG × 7'-9" TO 7-4 WIDE
TUBES	40 LARGE 5¼" O.D. × 7 S.W.G. 136 SMALL 2⅛" O.D. × 11 S.W.G.
SUPERHEATER ELEMENTS	1⅜" O.D. × 10 S.W.G.
LENGTH BETWEEN TUBEPLATES	17'-0"
HEATING SURFACES : TUBES	2264 SQ. FT.
FIREBOX	210 SQ. FT.
TOTAL EVAPORATIVE	2474 SQ. FT.
SUPERHEATER	718 SQ. FT.
TOTAL	3192 SQ. FT.
GRATE AREA	42 SQ. FT.

CYLINDERS (TWO)	19½" × 28"
TRACTIVE EFFORT	35912 LBS.
ADHESION FACTOR	4·22
BRAKE % ENGINE & TENDER (WITH TRAILING TRUCK BRAKED)	69·2 %
BRAKE LOAD ON ENGINE & TENDER (WITH TRAILING TRUCK BRAKED)	92·85 TONS.

WEIGHTS	FULL	EMPTY
ENGINE	94-12	85-15
TENDER	47-4	21-4
TOTAL	141-16	106-19

PROPOSED CLASS 8 2-8-2 ENGINE.

Fig. 5.

having 36 sq. ft. of grate area. By this means only a single design of bar frame would be required for the proposed Pacifics. The Class 5 then became a 4–6–0 having similar cylinders, valve gear and wheels as the Class 6, but carrying the excellent boiler of the L.M.S. Class 5 engine, as progressively developed under Stanier and Ivatt.

(3) Because of the very large numbers of 2–8–0s, both L.M.S. and W.D. which had been released to the railways after the war, there was no immediate need for the 2–8–2 freight engine. When that need did arise however it was coupled with the wish to obtain a higher performance than the existing 2–8–0s could provide, particularly as regards running speeds now limited by coupled wheels in the region of 4-ft. 8-in. diameter wheels and Fig. 5 indicates the proposed form it next took. By now the Class 7 4–6–2 had already been designed, and the new scheme included as many as possible of the features of the eventual "Britannia". Riddles was however a little unhappy about this solution, for he could not help observing that freight locomotive performance depended mainly upon adhesion, and that the increase in adhesion which the 2–8–2 was offering over the existing 2–8–0 was small in relation to the increase in potential power. Moreover his experience in designing and running the W.D. locomotives had impressed upon him the great merits of ten-coupled wheels for such duties. As soon therefore as some further scheming assured him that a 5-ft. 0-in. diameter coupled wheel could satisfactorily be incorporated under a wide firebox within the L_2 loading gauge, he opted for this combination in the form of a 2–10–0. The boiler could no longer be standard with that on the Pacifics, but all other components could still be standardised to a high degree.

The first group of standard locomotive thus became

Class 7	4–6–2	Mixed Traffic	70000
,, 6	4–6–2	Mixed Traffic	72000
,, 5	4–6–0	Mixed Traffic	73000
,, 9	2–10–0	Freight	92000

The development of all the components and details on these locomotives is set out in the next chapter, but the final aspect and leading dimensions of each are shown in Figs. 6-9 and Plates 7-10.

The second group of four further proposed designs were originally intended to be developments of existing L.M.S. types adapted to include some of the major features of the first group of entirely new locomotives, or alternatively to adapt an existing design to suit another power class.

Fig. 10 illustrates the proposals as set out in the report, but unlike the case of the first group, only one was radically altered in its whole conception before production, the other three being built largely as now outlined, subject to modification and development of details as the designs progressed.

The first of these engines was to deal with the heaviest and fastest passenger duties in new Class 8, and it will be seen that it was to be closely based upon the L.M.R. "Duchess" Class 4–6–2. It was pointed out that the existing boiler on this engine had an exceptionally successful

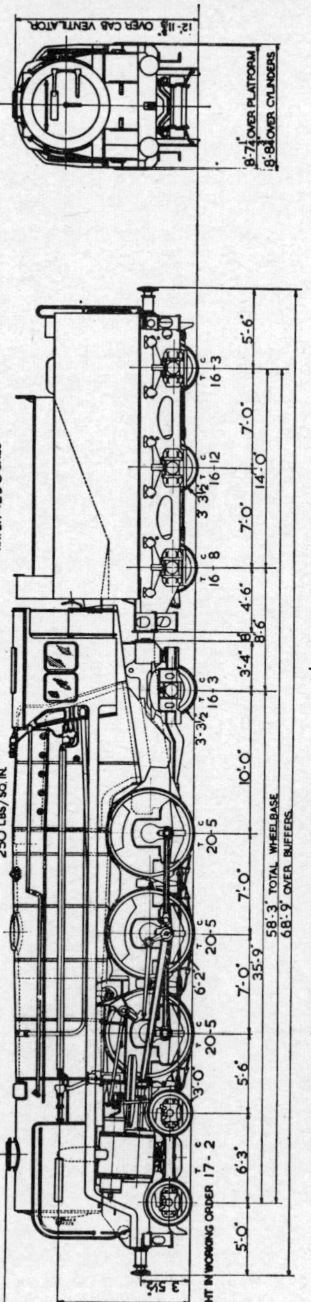

MINIMUM RADIUS CURVE WITHOUT GAUGE WIDENING 6 CHAINS (4½ CHAINS DEAD SLOW.)

BOILER BARREL DIAMETER (OUTSIDE)	5'-9" INCREASING TO 6'-5½"
FIREBOX (OUTSIDE)	7'-0" LONG × 7'-9" TO 7'-4" WIDE
TUBES	40 LARGE 5½" O.D. × 7 S.W.G.
	136 SMALL 2⅛" O.D. × 11 S.W.G.
SUPERHEATER ELEMENTS	1⅜" O.D. × 9 S.W.G.
LENGTH BETWEEN TUBEPLATES	17'-0"
HEATING SURFACES: TUBES	2264 SQ. FT.
FIREBOX	210 SQ. FT.
TOTAL EVAPORATIVE	2474 SQ. FT.
SUPERHEATER	677 SQ. FT.
FREE FLUE AREA.	6·8 SQ. FT.
GRATE AREA.	42 SQ. FT.

CYLINDERS - TWO	20" × 28"
TRACTIVE EFFORT	32,150 LBS
ADHESION FACTOR	4·23
BRAKE % ENGINE & TENDER	53·4.
BOILER TYPE	BR. 1
TENDER TYPE	BR. 1

WEIGHTS	FULL T C	EMPTY T C
ENGINE	94-0	85-3
TENDER	49-3	23-3
TOTAL	143-3	108-6

SCALE 0 2 4 6 8 10 FEET

STANDARD CLASS 7, 4-6-2 ENGINE.

Fig. 6.

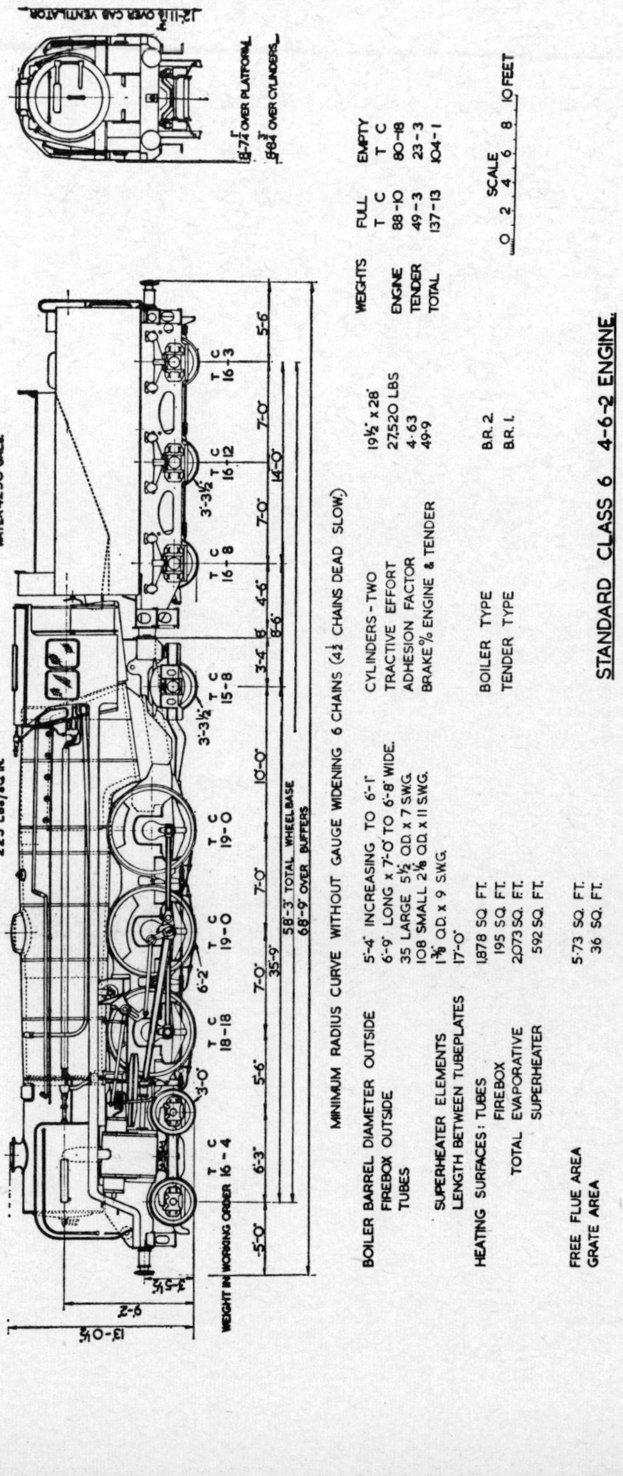
12'-11⅛" OVER CAB VENTILATOR
8'-7¼" OVER PLATFORM
8'-8¾" OVER CYLINDERS
225 LBS/SQ. IN.
COAL 7 TONS
WATER 4250 GALS.
13'-0½"
9'-2"
3'-5½"
WEIGHT IN WORKING ORDER
T C 16 - 4
T C 18-18
T C 19-0
T C 19-0
T C 15-8
T C 16-8
T C 16-12
T C 16-3
3'-0"
6'-2"
3'-3½"
3'-3½"
-5'-0"
6'-3"
5'-6"
7'-0"
7'-0"
10'-0"
3'-4
8'
4'-6"
7'-0"
7'-0"
5'-6"
8'-6"
14'-0"
35'-9"
58'-3" TOTAL WHEELBASE
68'-9" OVER BUFFERS
MINIMUM RADIUS CURVE WITHOUT GAUGE WIDENING 6 CHAINS (4½ CHAINS DEAD SLOW.)
BOILER BARREL DIAMETER OUTSIDE 5'-4" INCREASING TO 6'-1"
FIREBOX OUTSIDE 6'-9" LONG x 7'-0" TO 6'-8" WIDE.
TUBES 35 LARGE 5½ O.D. x 7 S.W.G.
108 SMALL 2⅛ O.D. x 11 S.W.G.
SUPERHEATER ELEMENTS 1⅜" O.D. x 9 S.W.G.
LENGTH BETWEEN TUBEPLATES 17'-0"
HEATING SURFACES: TUBES 1,878 SQ. FT.
FIREBOX 195 SQ. FT.
TOTAL EVAPORATIVE 2,073 SQ. FT.
SUPERHEATER 592 SQ. FT.
FREE FLUE AREA 5·73 SQ. FT.
GRATE AREA 36 SQ. FT.
CYLINDERS - TWO 19½" x 28"
TRACTIVE EFFORT 27,520 LBS
ADHESION FACTOR 4·63
BRAKE % ENGINE & TENDER 49·9
BOILER TYPE B.R. 2
TENDER TYPE B.R. 1
WEIGHTS FULL T C EMPTY T C
ENGINE 88-10 80-18
TENDER 49-3 23-3
TOTAL 137-13 104-1
SCALE
0 2 4 6 8 10 FEET
STANDARD CLASS 6 4-6-2 ENGINE.

Fig. 7.

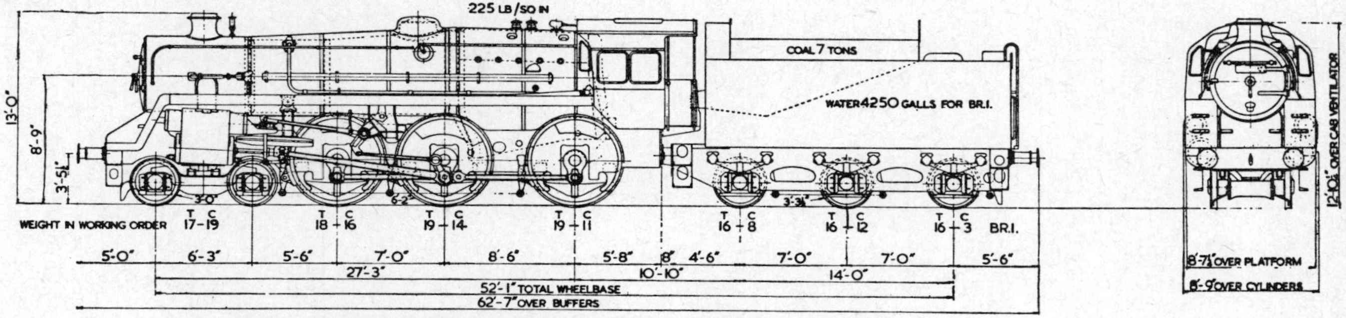

MINIMUM RADIUS CURVE WITHOUT GAUGE WIDENING 6 CHAINS (4½ CHAINS DEAD SLOW).

BOILER BARREL DIAMETER (OUTSIDE)	4'-11 13/16" INCREASING TO 5'-8½"	CYLINDERS (TWO)	19" x 28"
FIREBOX (OUTSIDE)	9'-2 13/16" LONG x 3'-11¼" WIDE	TRACTIVE EFFORT	26120 LB
TUBES	28 LARGE 5⅛" O.D. x 7 SWG	ADHESION FACTOR	4.97
	151 SMALL 1⅞" O.D. x 11 SWG	BRAKE % ENGINE & TENDER	60.7 BR.I. TENDER
SUPERHEATER ELEMENTS	1⅜" O.D. x 9 SWG		
LENGTH BETWEEN TUBEPLATES	13'-2⅞"		
HEATING SURFACES TUBES	1479 SQ.FT.	BOILER TYPE	BR.3.
FIREBOX	171 SQ.FT.	TENDER TYPE	BR.I.
TOTAL EVAPORATIVE	1650 SQ.FT.		
SUPERHEATER	358 SQ.FT.		
FREE FLUE AREA	4.5 SQ.FT.		
GRATE AREA	28.7 SQ.FT.		

WITH BR.I. TENDER

WEIGHTS	FULL T–C	EMPTY T–C
ENGINE	76–0	69–5
TENDER	49–3	23–3
TOTAL	125–3	92–8

SCALE
0 2 4 6 8 10 FEET

STANDARD CLASS 5. 4-6-0 ENGINE

Fig. 8.

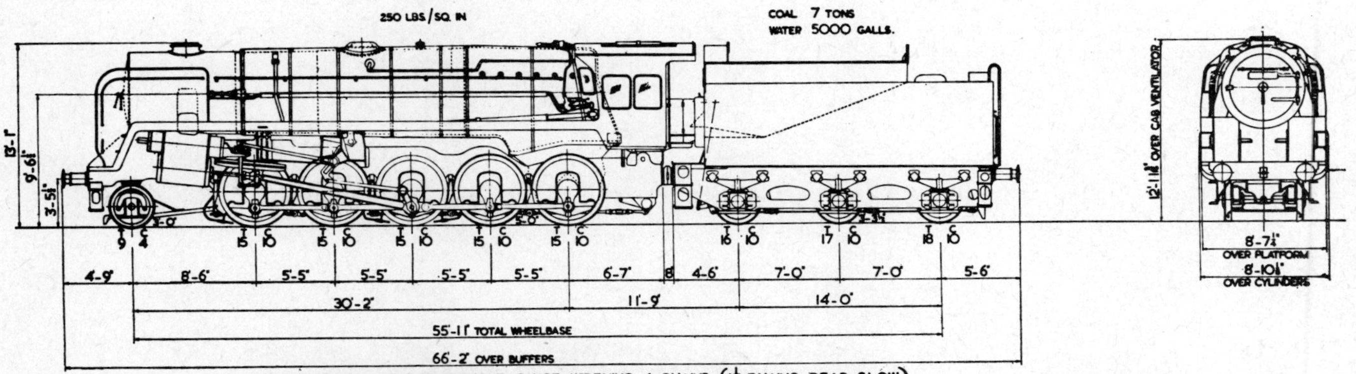

MINIMUM RADIUS CURVE WITHOUT GAUGE WIDENING 6 CHAINS (4½ CHAINS DEAD SLOW.)

BOILER BARREL DIAMETER (OUTSIDE)	5'-9" INCREASING TO 6'-1"
FIREBOX (OUTSIDE)	7'-5¼" LONG x 7'-0⅞" TO 6'-6⅞" WIDE
TUBES	35 LARGE 5¼" O.D. x 7 S.W.G.
	138 SMALL 2" O.D. x 11 S.W.G.
SUPERHEATER ELEMENTS	1⅜" O.D. x 9 S.W.G.
LENGTH BETWEEN TUBEPLATES	15'-3"
HEATING SURFACES: TUBES	1836 SQ. FT.
FIREBOX	179 SQ. FT.
TOTAL EVAPORATIVE	2015 SQ. FT.
SUPERHEATER	535 SQ. FT.
FREE FLUE AREA	5·49 SQ. FT.
GRATE AREA	40·2 SQ. FT.

CYLINDERS (TWO)	20" x 28"
TRACTIVE EFFORT	39,667 LBS.
ADHESION FACTOR	4·38
BRAKE % ENGINE & TENDER	69·8
BOILER TYPE	B.R. 9
TENDER TYPE	B.R. 1G

WEIGHTS	FULL T	FULL C	EMPTY T	EMPTY C
ENGINE	86	14	78	0
TENDER	52	10	23	3
TOTAL	139	4	101	3

SCALE 0 2 4 6 8 10 FEET

B.R. STANDARD CLASS 9. 2-10-0 ENGINE

Fig. 9.

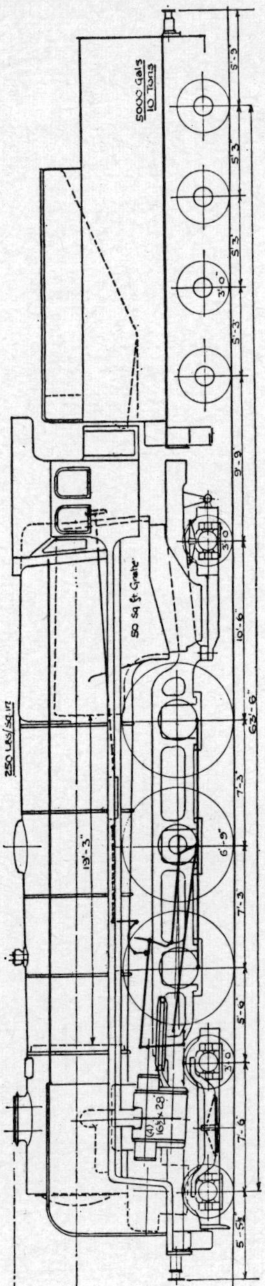

Fig. 10. Original proposals for 4–6–2, 4–6–0, and 2–6–0 locomotives. Above: Type 75 (passenger) class 7 with 22½ton axle-loading; below: Type 76 (mixed traffic) class 4 with 17ton axle-loading (top) and Type 78 (mixed traffic) class 3 with 16ton axle-loading (bottom).

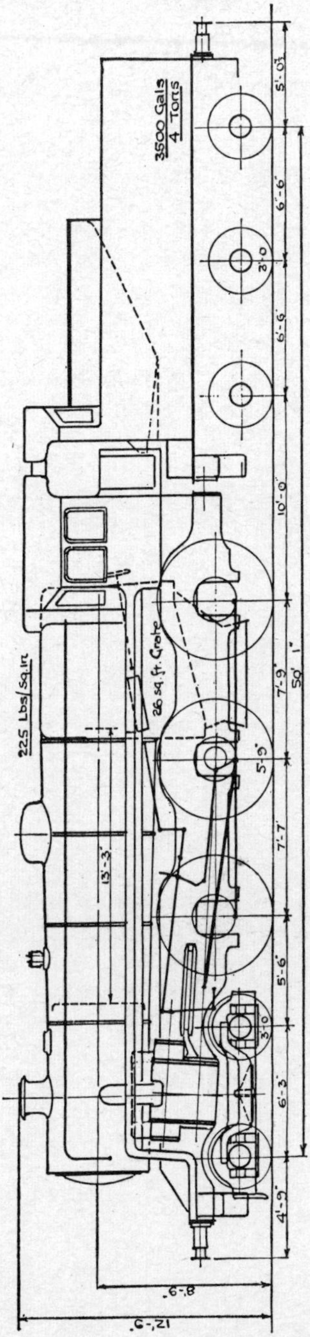
3500 Gals
4 Tons
225 Lbs/sq.in
26 sq. ft. Grate
13'-3"
50'-1"
5'-0½"
6'-6"
6'-6"
10'-0"
7'-9"
7'-7"
5'-6"
6'-3"
4'-9"
3'-0"
5'-9"
8'-9"
12'-9"

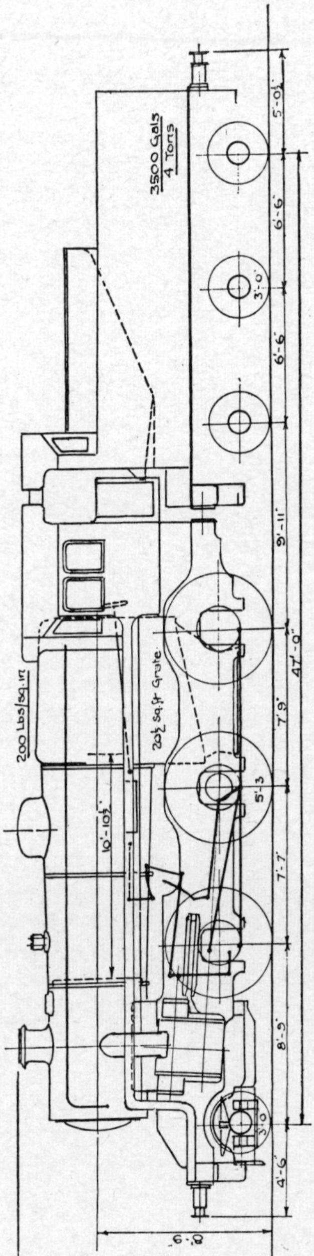
3500 Gals
4 Tons
200 Lbs/sq.in
20½ sq. ft. Grate
10'-10½"
47'-9"
5'-0½"
6'-6"
6'-6"
9'-11"
7'-9"
7'-7"
8'-9"
4'-6"
3'-0"
5'-3"
8'-9"
12'-9"

maintenance record, together with good steaming qualities and that no good purpose would be served by redesigning it. Since two cylinders would not longer suffice within the loading gauge, the four-cylinder arrangement of the Duchess was preferred to the three-cylinders of the E.N.E. A4 class for example, notwithstanding that it did not represent the least possible number of cylinders to do the job. The report pointed out that in 1941, and although admittedly under wartime conditions, 652 L.N.E.R. three-cylinder engines had experienced ten times as many hot bearings on inside as on outside big ends, and the still current practice of fitting stink bombs to give warning when inside big ends were getting warm was a measure of the anxiety which was experienced. Reference was also made to 189 three-cylinder engines of former Class 5X on the L.M.S. which averaged 24 inside big end failures per annum over a seven-year period—many of them involving fracture of the inside cylinder and very expensive repairs. Here again the unlikable stink bombs had been introduced. From this kind of experience, coupled with the fact that for a given total power, four-cylinders permitted lower individual loads on the inside big ends, and also because of the very simple and robust rocker mechanism by which four valves can be driven by two sets of valve gear, for all these reasons, a four-cylinder arrangement was supported.

The main modifications proposed for the "Duchess" were the provision of bar frames, use of a larger ashpan and tender, and the widest use of standard fittings and details. Here again, there was no immediate demand for such a locomotive, each region for the time being having amply replenished its stock of Class 8 engines with products to its own design. Indeed no such need ever did arise before steam duties were overtaken by the onset of Diesel Traction.

An unexpectedly circumstance however provided the opportunity to have a single Class 8 locomotive built as a prototype, which was eventually developed along very different lines from those indicated above. In October 1952 the Harrow train disaster completely destroyed L.M.R. Pacific No. 46202 *Princess Anne* only a few weeks after her conversion from Stanier's "Turbomotive" into a conventional four-cylinder engine. The vacancy in the stock list thus created enabled Riddles to seek authority for a replacement in the form of a B.R. Class 8 4–6–2 on which running experience might be gained against possible future needs for more of this kind of locomotive.

Bear in mind that, although we now have afterknowledge of the pace of dieselisation, it was by no means certain then, nor for some time threeafter, that more Class 8 steam engines might not eventually be required.

By this time much water had flowed under the bridge, bar frames were out for reasons described in a subsquent chapter and the Class 7 "Britannias" were already at work.

There emerged from all this a complete re-thinking of the Class 8 project which went into production design with three-cylinders, Improved Caprotti Valve Gear, a "Britannia" boiler with enlarged grate and a

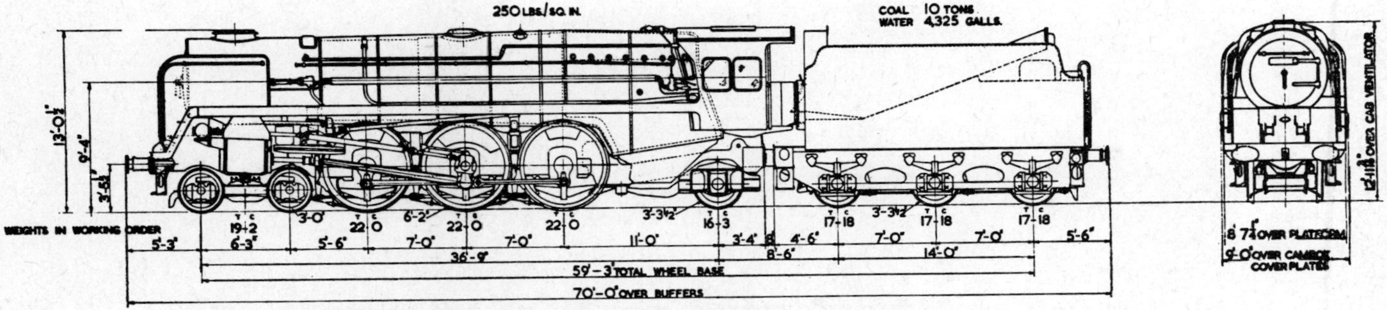

MINIMUM RADIUS CURVE WITHOUT GAUGE WIDENNG 6 CHAINS (4½ CHAINS DEAD SLOW)

BOILER BARREL DIAMETER (OUTSIDE)	5'-9" INCREASING TO 6'-5½"
FIREBOX (OUTSIDE)	8'-0⅛" LONG x 7'-9" TO 7'-4" WIDE
TUBES	40 LARGE 5¼" O.D. x 7 S.W.G.
	136 SMALL 2⅛" O.D. x 11 S.W.G.
SUPERHEATER ELEMENTS	1⅜" O.D. x 9 S.W.G.
LENGTH BETWEEN TUBEPLATES	17'-0"
HEATING SURFACES: TUBES	2264 SQ. FT.
FIREBOX	226 SQ. FT.
TOTAL EVAPORATIVE	2490 SQ. FT.
SUPERHEATER	691 SQ. FT.
FREE FLUE AREA	6·8 SQ. FT.
GRATE AREA	48·6 SQ. FT.

CYLINDERS (THREE)	18" x 28"
TRACTIVE EFFORT	39,080 LBS.
ADHESION FACTOR	3·78
BRAKE % ENGINE & TENDER	51·17
BOILER TYPE	B.R. 13
TENDER TYPE	B.R. 1J.

WEIGHTS	FULL	EMPTY
	T. C.	T. C.
ENGINE	101-5	92-0
TENDER	53-14	24-8
TOTAL	154-19	116-8

SCALE 0 2 4 6 8 10

STANDARD CLASS 8. 4-6-2 ENGINE.
BRITISH-CAPROTTI VALVE GEAR.

Fig. 11.

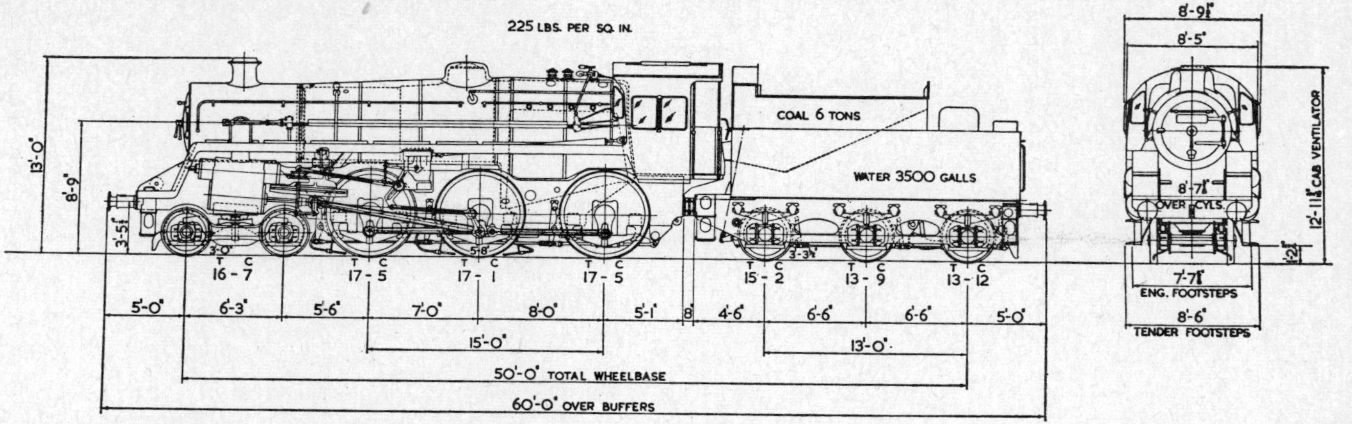

BOILER BARREL DIAMETER (OUTSIDE)	4'-9" INCREASING TO 5'-3"
FIREBOX (OUTSIDE)	8'-6" LONG x 4'-0⅛" WIDE
TUBES	21 LARGE 5⅛" O.D. x 7 S.W.G.
	157 SMALL 1⅞" O.D. x 12 S.W.G.
SUPERHEATER ELEMENTS	1⅜" O.D. x 9 S.W.G.
LENGTH BETWEEN TUBEPLATES	13'-0"
HEATING SURFACES: TUBES	1301 SQ. FT.
FIREBOX	143 SQ. FT.
TOTAL EVAPORATIVE	1444 SQ. FT.
SUPERHEATER	258 SQ. FT.
FREE FLUE AREA	3·78 SQ. FT.
GRATE AREA	26·7 SQ. FT.

CYLINDERS (TWO)	18" x 28"
TRACTIVE EFFORT	25,515 LBS.
ADHESION FACTOR	4·52
BRAKE % ENGINE & TENDER	55·1
MINIMUM RADIUS CURVE WITHOUT GAUGE WIDENING.	6 CHAINS (OR 4½ CHAINS DEAD SLOW)
BOILER TYPE	BR 4
TENDER TYPE	BR 2

WEIGHTS	FULL T C	EMPTY T C
ENGINE	67-18	63-6
TENDER	42-3	20-10
TOTAL	110-1	83-16

SCALE 0 2 4 6 8 10 FEET

B. R. STANDARD CLASS 4. 4-6-0 ENGINE.

Fig. 12.

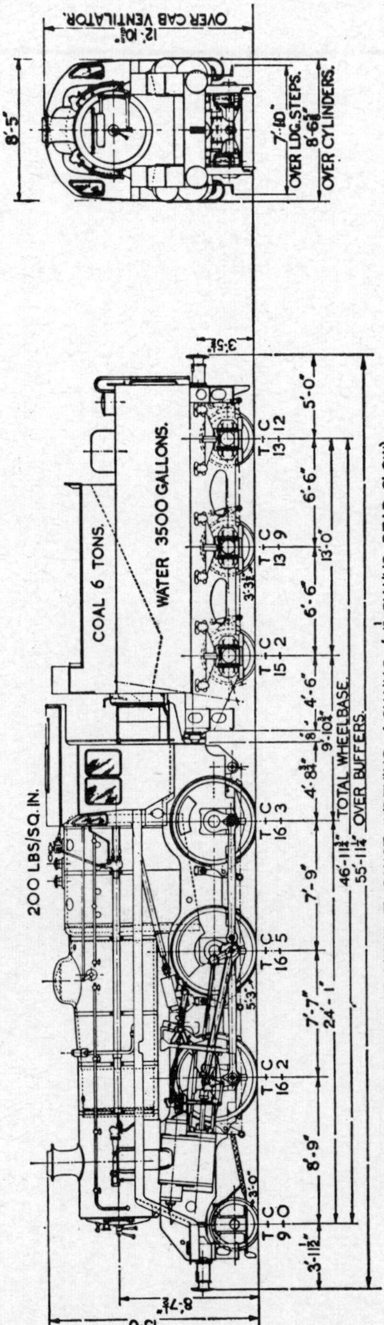

MINIMUM RADIUS CURVE WITHOUT GAUGE WIDENING 6 CHAINS (4½ CHAINS DEAD SLOW).

BOILER BARREL DIAMETER (OUTSIDE)	4'-5" INCREASING TO 5'-0½"
FIREBOX (OUTSIDE)	7'-0" LONG X 4'-0" WIDE.
TUBES	18 LARGE 5⅛" O.D. X 7 S.W.G.
	143 SMALL 1⅝" O.D. X 12 S.W.G.
SUPERHEATER ELEMENTS	1 3/16" O.D. 9 S.W.G.
LENGTH BETWEEN TUBEPLATES	10'-10½"
HEATING SURFACES TUBES	923·54 SQ. FT.
FIREBOX	118·42 " "
TOTAL EVAPORATIVE	1041·96 " "
SUPERHEATER	184·50 " "
FREE FLUE AREA	3·054 " "
GRATE AREA	20·35 " "

CYLINDERS (TWO)	17½" X 26".
TRACTIVE EFFORT	21490 LBS.
ADHESION FACTOR	5·05
BRAKE % ENGINE & TENDER	60·7
BOILER TYPE	BR. 6.
TENDER TYPE	BR. 2 A.

WEIGHTS	FULL	EMPTY
	T - C	T - C
ENGINE	57-10	53-0
TENDER	42-3	20-10
TOTAL	99-13	73-10

SCALE
0 1 2 3 4 5 6 7 8 9 10

Fig. 13.

STANDARD CLASS 3. 2-6-0 ENGINE.

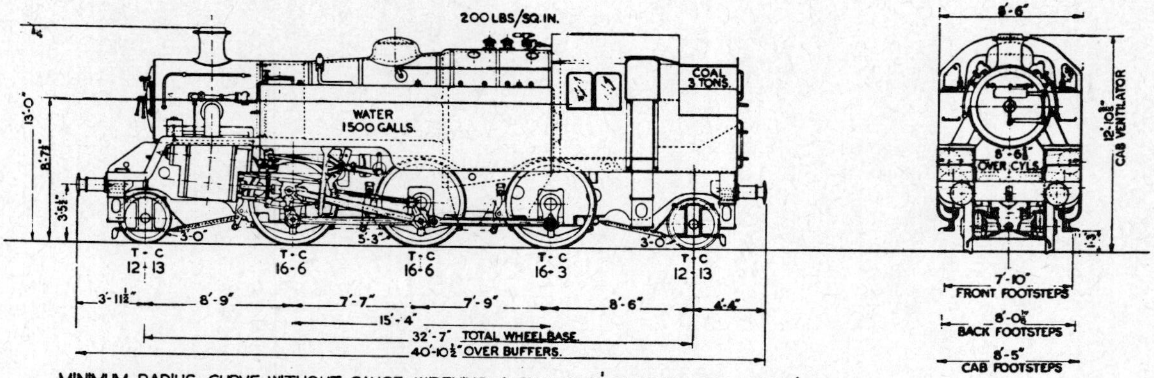

MINIMUM RADIUS CURVE WITHOUT GAUGE WIDENING 6 CHAINS (4½ CHAINS DEAD SLOW).

BOILER BARREL DIAMETER (OUTSIDE)	4'-5" INCREASING TO 5'-0½".
FIREBOX (OUTSIDE)	7'-0" LONG X 4'-0" WIDE.
TUBES	18 LARGE 5⅛" O.D. x 7 S.W.G.
	143 SMALL 1⅝" O.D. x 12 S.W.G.
SUPERHEATER ELEMENTS	1⅜" O.D. 9 S.W.G.
LENGTH BETWEEN TUBEPLATES	10'-10½".
HEATING SURFACES. TUBES	923·54 SQ.FT
FIREBOX	118·42 " "
TOTAL EVAPORATIVE	1041·96 " "
SUPERHEATER	184·50 " "
FREE FLUE AREA	3·054 " "
GRATE AREA	20·35 " "

CYLINDERS (TWO)	17½" x 26".
TRACTIVE EFFORT	21490 LBS.
ADHESION FACTOR	5·08
BRAKE % ENGINE	51·5
BOILER TYPE	BR 6.

WEIGHTS.	FULL.	EMPTY.
	T-C	T-C
	74-1.	60-18.

SCALE
0 2 4 6 8 10 FEET

B R. STANDARD CLASS 3. 2-6-2 TANK ENGINE.

Fig. 14.

super inside big end design, all of which will be described in due course—the eventual No. 71000 *Duke of Gloucester* Fig. 11 gives the leading dimensions and Plate 11 shows the engine as built.

The Class 4 4–6–0 was proposed as a development of the L.M.R. Class 4 2–6–4 tank, an engine which had won a high reputation for speed, reliability, and general effectiveness. There were areas in the country however—Central Wales was a case in point—which called for a longer working range than a tank engine could provide, but where lower axle loads were called for than those of Class 5 engines. The W.R. 78XX "Manor" class was just such an engine, but its width over cylinders precluded universal running. This locomotive as designed for production varied only in details from Fig. 10, and with a 17-ton axleload, it satisfied the L_1 loading gauge which meant that it could run almost anywhere, and it was largely assembled from existing L.M.R. flanging blocks and B.R. standard details.

The two Class 3 designs, 2–6–0 and corresponding 2–6–2 tank arose from the existence at the time of nationalisation of a number of routes having a 16-ton axle load restriction which Class 4 engines could not satisfy. There was no existing L.M.R. boiler in this category, or at least none that was sufficiently satisfactory, and it was decided eventually to use the flanging blocks of the Swindon 5100 class tank engines, but in all other respects the locomotives were built as first diagrammed, conforming to the general standardisation in all their details.

Thus the second group was finally produced as follows:—

Class 8	4–6–2	Express Passenger	71000
„ 4	4–6–0	Mixed Traffic	75000
„ 3	2–6–0	Mixed Traffic	77000
„ 3	2–6–2T	Mixed Traffic	82000

Figs. 11, 12, 13 and 14 give the dimensions of the above and Plates 11, 12, 13 and 14 show the finished product.

The last group of four types, making up the total of 12, were at first intended to be existing L.M.R. designs modified only to include the standard details. These were the three Ivatt engines, Class 4 and Class 2 2–6–0 and Class 2 2–6–2T, and the Fairburn 2–6–4T. These locomotives already included all of the design features for easier maintenance which it was desired to introduce, they were first-class performers in their different power categories and their low axle loads gave them the required route availability. They were already accepted for wide use outside their parent system on the Eastern, North Eastern and Southern Regions, but for still wider use the 2–6–4 tank engine required fairly drastic alterations to cylinders and superstructure and adoption of a higher boiler pressure in order to enable it to satisfy the universal L_1 loading gauge. The others were only altered so far as was required to accept the standard fittings, and to incorporate the improvements in blastpipe and chimney proportions which resulted from a series of Swindon tests with the original L.M.R. types. The final group was thus:—

Class 4 2–6–0 Mixed Traffic 76000

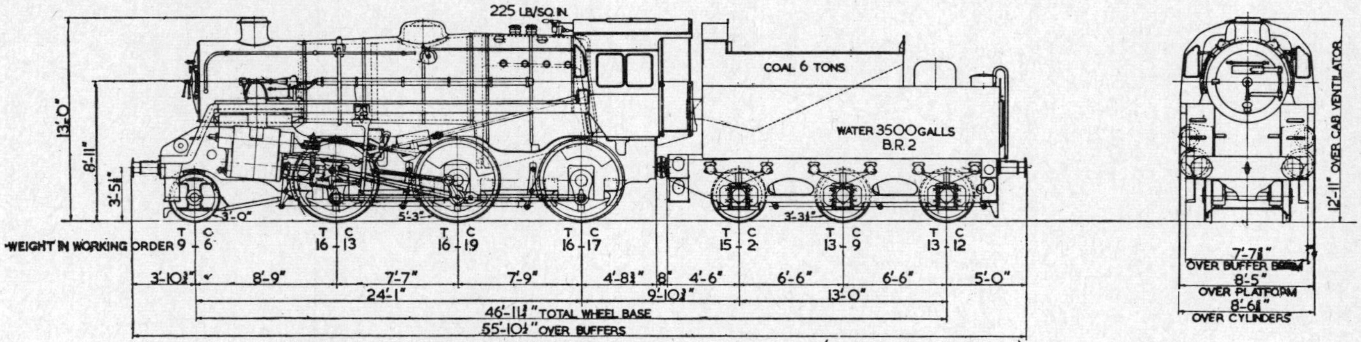

RADIUS OF MINIMUM CURVE. 6 CHAINS. WITHOUT GAUGE WIDENING (4½ CHAINS DEAD SLOW)

BOILER BARREL DIAMETER (OUTSIDE)	4'-9½" INCREASING TO 5'-3"
FIREBOX (OUTSIDE)	7'-6" LONG x 4'-0½" WIDE
TUBES	24 LARGE 5⅛" O.D. x 7 S.W.G.
	156 SMALL 1⅝" O.D. x 12 S.W.G.
SUPERHEATER ELEMENTS	1⅜" O.D. x 9 S.W.G.
LENGTH BETWEEN TUBEPLATES	10'-10½"
HEATING SURFACES TUBES	1,075 SQ.FT.
FIREBOX	131 SQ.FT.
TOTAL EVAPORATIVE	1,206 SQ.FT.
SUPERHEATER	247 SQ.FT.
FREE FLUE AREA	3.8 SQ.FT.
GRATE AREA	23 SQ.FT.

CYLINDERS (TWO)	17½" x 26"
TRACTIVE EFFORT	24,170 LB.
ADHESION FACTOR	4.68
BRAKE % ENGINE & TENDER	63.75
BOILER TYPE	BR 7.
TENDER TYPE	BR 2.

WEIGHTS	FULL T C Q	EMPTY T C Q
ENGINE	59-15-0	55-18-2
TENDER	42-3-0	20-10-2
TOTAL	101-18-0	76-9-0

SCALE 0 1 2 3 4 5 6 7 8 9 10 FEET

STANDARD CLASS 4. 2-6-0 ENGINE

Fig. 15.

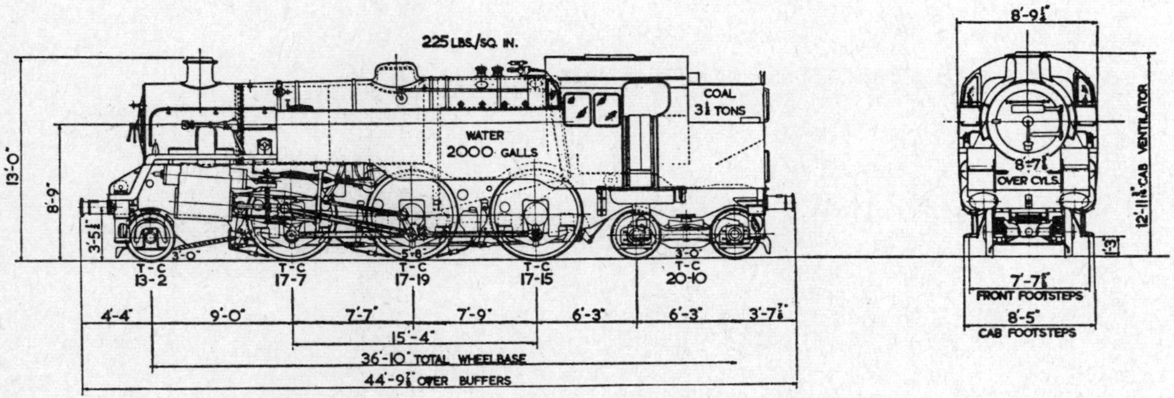

BOILER BARREL DIAMETER (OUTSIDE)	4'-9" INCREASING TO 5'-3"
FIREBOX (OUTSIDE)	8'-6" LONG × 4'-0½" WIDE
TUBES	21 LARGE 5⅛" O.D. × 7 S.W.G. 157 SMALL 1¾" O.D. × 12 S.W.G.
SUPERHEATER ELEMENTS	1⅜" O.D. × 9 S.W.G.
LENGTH BETWEEN TUBEPLATES	12'-3"
HEATING SURFACES: TUBES	1223 SQ. FT.
FIREBOX	143 SQ. FT.
TOTAL EVAPORATIVE	1366 SQ. FT.
SUPERHEATER	240 SQ. FT.
FREE FLUE AREA	3·78 SQ. FT.
GRATE AREA	26·7 SQ. FT.

CYLINDERS (TWO)	18" × 28"
TRACTIVE EFFORT	25,515 LBS.
ADHESION FACTOR	4·65
BRAKE %	44·2
MINIMUM RADIUS CURVE WITHOUT GAUGE WIDENING	6 CHAINS (OR 4½ CHAINS DEAD SLOW)
BOILER TYPE	BR 5

WEIGHTS	FULL	EMPTY
	T C	T C
	86-13	69-8

SCALE
0 2 4 6 8 10 FEET

B.R. STANDARD CLASS 4. 2-6-4 TANK ENGINE

Fig. 16.

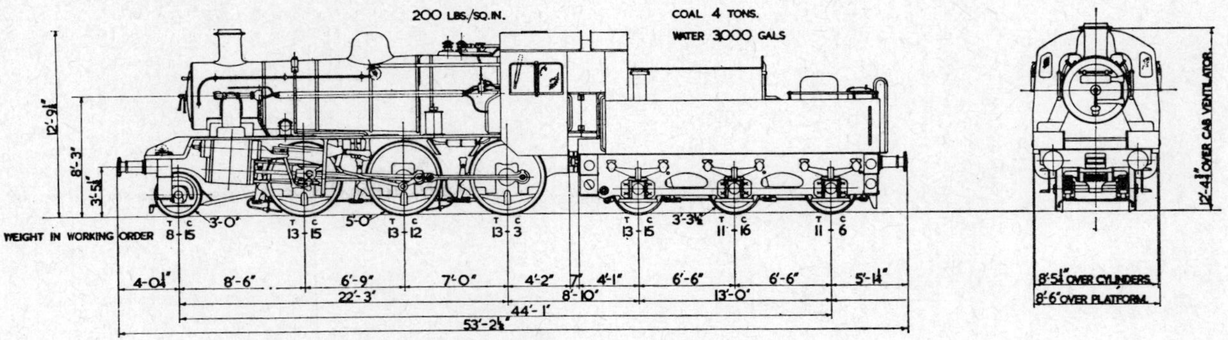

MINIMUM RADIUS CURVE WITHOUT GAUGE WIDENING 4 CHAINS.

BOILER BARREL DIAMETER (OUTSIDE)	4'-3" INCREASING TO 4'-8"
FIREBOX (OUTSIDE)	5'-11" LONG x 4'-0 7/16" WIDE
TUBES	12 LARGE 5⅛" O.D. x 7 S.W.G.
	162 SMALL 1⅝" O.D. x 12 S.W.G.
SUPERHEATER ELEMENTS	1⅜" O.D.
LENGTH BETWEEN TUBEPLATES	10'-10½"
HEATING SURFACES: TUBES	924 SQ. FT.
FIREBOX	101 SQ. FT.
TOTAL EVAPORATIVE	1025 SQ. FT.
SUPERHEATER	124 SQ. FT.
FREE FLUE AREA	2·77 SQ. FT.
GRATE AREA	17·5 SQ. FT.

CYLINDERS (TWO)	16½" DIA. x 24" STROKE
TRACTIVE EFFORT	18,513 LBS.
ADHESION FACTOR	4·9
BRAKE % ENGINE & TENDER	60·23
BOILER TYPE	B.R. 8
TENDER TYPE	B.R. 3

WEIGHTS	FULL T C	EMPTY T C
ENGINE	49-5	45-8
TENDER	36-17	19-9
TOTAL	86-2	64-17

SCALE
0 2 4 6 8 10 FEET

STANDARD CLASS 2. 2-6-0 ENGINE.

Fig. 17.

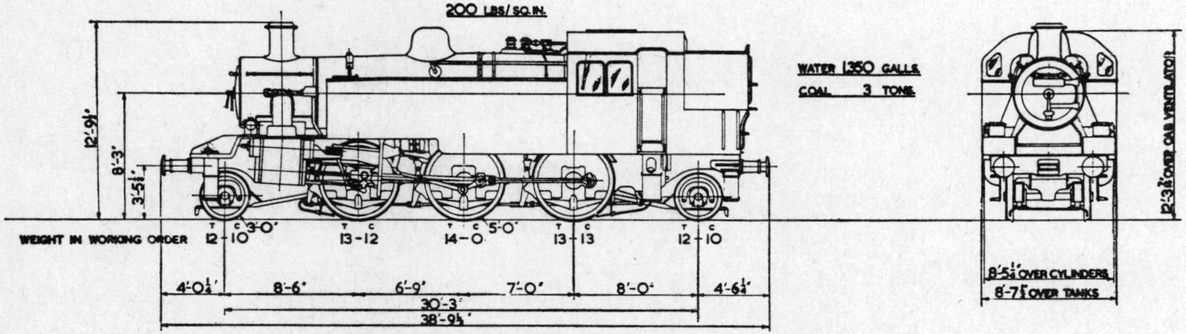

MINIMUM RADIUS CURVE WITHOUT GAUGE WIDENING 4 CHAINS.

BOILER BARREL DIAMETER (OUTSIDE)	4-3 INCREASING TO 4-8'
FIREBOX (OUTSIDE)	5-11 LONG x 4'-0 7/16" WIDE
TUBES	12 LARGE 5⅛" O.D. x 7 S.W.G.
	162 SMALL 1⅝" O.D. x 12 S.W.G.
SUPERHEATER ELEMENTS	1 3/16" O.D.
LENGTH BETWEEN TUBEPLATES	10'-10½"
HEATING SURFACES:- TUBES	924 SQ. FT.
FIREBOX	101 SQ. FT.
TOTAL EVAPORATIVE	1025 SQ. FT.
SUPERHEATER	124 SQ. FT.
FREE FLUE AREA	2·77 SQ. FT.
GRATE AREA	17·5 SQ. FT.

CYLINDERS (TWO)	16½" DIA. x 24" STROKE
TRACTIVE EFFORT	18,513 LBS.
ADHESION FACTOR	5·0
BRAKE %	47
BOILER TYPE	B.R. 8

WEIGHTS	FULL	EMPTY
	T C	T C
ENGINE	66-5	53-12

SCALE
0 2 4 6 8 10 FEET.

STANDARD CLASS 2. 2-6-2 TANK ENGINE.

Fig. 18.

Class 4 2–6–4T Mixed Traffic 80000
,, 2 2–6–0 Mixed Traffic 78000
,, 2 2–6–2T Mixed Traffic 84000

Figs. 15-18 give the dimensions of each class, and Plates 15-18 indicate the appearance.

Thus were the 12 B.R. standard types fashioned out of personalities and circumstances. In a final and possibly unnecessary apologia for the "L.M.S." thinking which suffused the whole pattern, the first report contained these words in conclusion:—

"No standardisation scheme would possibly include the designs of all four regions.

No case can be made for preparing entirely new designs simply to avoid following one particular region.

The modern standard designs of each region have been developed to a high degree. No choice between them seems possible on the grounds of pure merit, as indicated in the first report of the Locomotive Standards Committee. An arbitrary decision has accordingly to be made on other grounds.

It so happens that in certain cases, L.M. designs have the highest route availability for their power, and at the same time incorporate the latest advances in design from an operating and maintenance point of view.

In considering possible alternatives, Western Region designs are for the most part ruled out of general standardisation due to the effect of their width over cylinders on route availability. The E.N.E. types tend to be heavier and have a more restricted route availability than their L.M. counterparts, and the latest S.R. engines do not provide the requirements of simplicity and ease of maintenance which is the cornerstone of the present scheme."

So much then for the inception and ultimate realisation of the locomotives as a whole. Let us now in the next chapter consider what occured between the first thoughts and the end product—how the engines were designed, and how the main components and the fittings were selected.

IV

The Design Process and Development of the Boilers

THE RECOMMENDATIONS of the first report of June 1948 on the proposed standard locomotives were, after digestion by the Railway Executive, considered by the regions and by the "ideal stock" committee with the result that ten out of the 12 steam types were accepted for earliest inclusion in future building programmes in substitution for regional designs. Only the largest Class 8 Pacific and the large freight engine were excluded at this time, and did not in fact appear on rail for another five to six years. By March 1949 requirements had been finalised, and production design work was initiated. A diagram was prepared at R.E. Headquarters setting out the locomotives as finally agreed, that is with Class 7 and 6 Pacifics and a 4–6–0 Class 5, the other types remaining as in the first report. This diagram RD10 was accompanied by a brief document explaining that the objective was to have as many of these ten types as possible included in the 1951 Building Programme, and that the Operating Department had indicated its first thoughts by issuing a priority for design consisting of, first, the Class 7 and 6 Pacifics, secondly the Class 4 4–6–0, then the Class 3 2–6–0 and 2–6–2T followed by the Class 5 4–6–0, the remaining types bringing up the rear. The document also contained an indication of the major component standardisation which it was hoped to achieve, six sizes of coupled wheels, five cylinder sizes, seven sets of flanging blocks for the boilers and so on. Finally, it summarised the work of the Locomotives Standards Committee, so far, on recommendations for standard fittings, half of which had been brought to the stage of draft specifications, the remaining half being still under consideration.

On March 24th, 1949, production design was launched at a meeting held at R.E. Headquarters with myself in the chair, and attended by the chief draughtsmen from Brighton, Derby, Doncaster and Swindon. The Class 7 and 6 Pacifics were allocated to Derby, the Class 5 4–6–0 to Doncaster, the Class 4 4–6–0 to Brighton and the Class 3 2–6–0 and 2–6–2T to Swindon. At later dates, the remaining design allocations were completed by giving the Class 8 4–6–2 to Derby, the Class 9 2–10–0 to Brighton, the Class 4 2–6–4 tank to Brighton, the Class 4 2–6–0 to Doncaster and the Class 2 2–6–0 and 2–6–2T to Derby. While each office undertook the overall design and the issue of drawings in respect of the engines allocated to them, they also undertook the design of certain components on behalf of the whole standard series. Thus Derby undertook bogies, two-axle trucks, wheels, axles, axleboxes, springs and tenders;

Doncaster did cylinders, slidebars, crossheads, coupling and connecting rods and valve gear; Brighton did sanding and brake gear, and details for pipe and rod and lubrication layouts; Swindon took over all boiler and ashpan mountings and details. Overlapping was minimised by identifying each of these headings with numbered sections in the Central Ordering Specification, the document already in wide use which extracted all components from the working drawings and listed them in a form suitable for materials ordering and workshop planning. By this means it was possible to control the aspect of standardisation, since only a single office was concerned in the working out and finalising drawings for any given detail. For each of these details there had of course been four or more separate versions under the former companies. In each case they required examination and discussion on their merits and upon their suitability for inclusion in the new locomotives. This was done at subsequent monthly meetings of the chief draughtsmen and on intermediate occasions when I visited each of the four offices within the same period and was able to discuss the progress of the work and give decisions at each draughtsman's board. While the control of the whole design process lay in my own hands, it was of course closely overlooked by Riddles, and the more difficult or controversial decisions were taken by him after bringing Bond in on the workshop aspect, or after reference to Operating and Motive power interests. Riddles was, however, an ideal chief to serve in that he never interfered with the main body of the work or with my responsibility for it, but on those occasions when his over-riding decision became necessary it was forthcoming at once, and in crisp and unambiguous terms.

For the chief draughtsmen and the staff in the four drawing offices who now embarked on this collective endeavour, it was a novel and interesting experience. Conditioned all their professional careers to uphold the practice of the companies to which they had owed allegiance, and subject respectively in the immediate past to the powerful "dictat" of the Stanier-Fairburn-Ivatt, Gresley-Thompson-Peppercorn, Collett-Hawkesworth, or Maunsell-Bulleid régimes, they were now required to turn their minds to an objective assessment of what they had done, and to see only part of it acceptable for future application. For the rest their minds had to broaden sufficiently to identify merit in hitherto alien practices, and to accept with good grace decisions based not necessarily upon any faults in what had previously been designed, but solely upon the need to apply one and not four versions of any given item. It was a wonderful shaking down process, calling for a measure of courage in the main participants for, in the first few years at any rate, all this work had to be accomplished under the eyes of the last of the former C.M.E.s who still remained on the premises, but were berefit of design responsibility. However, much it went against the grain, they stood aside from the standard locomotive design effort and, rather naturally, they were not able to approve much of it. Human nature being what it is, they could not deny themselves mordant asides to their former henchmen, who now had to steel themselves to serve new masters.

I cannot pay too high a tribute to the manner in which both officers and men in the different offices faced and accomplished their new tasks. Broadmindedness and good humour lubricated all our contacts, and goodwill and co-operation were forthcoming in ample measure. Indeed, I do not recall a single sour note during the whole of this hard slogging work in what was for many an unfamiliar medium. The Derby men had the easiest time of it psychologically, for most of them had already experienced a similar undertaking when Stanier came to the L.M.S., and the design philosophy of the new standard engines was very much in the Derby tradition.

In 1948, T. F. Coleman was in charge of design as a whole at Derby. Although he so remained until July 1949, he was not, in view of his impending retirement, too interested in concerning himself with the day to day detail for something of which he would neither see the continuation nor the end. He therefore left to his lieutenants most of the active participation, but he did take a fatherly interest in the whole proceedings, and contributed one or two suggestions of the greatest value as the work proceeded. J. W. Caldwell had been Chief Locomotive Draughtsman since 1945 and so continued until 1956. It was he who had had a lot to do with the form which the Ivatt Class 2 and Class 4 engines had taken, and far from showing any resistance to the somewhat American trend in the standard designs, he had sometimes to be gently restrained from a wish to see even more bizarre features introduced. At Doncaster, E. Windle reigned over the design office, a small but active man who had tasted of the asperities of the Thompson régime, but who now came to this work from the more mellow and benign influence of Peppercorn, last C.M.E. of the L.N.E. Railway. At Swindon F. C. Mattingly was a pure-bred Great Western man, nurtured in the unbroken traditions of that line, and uncontaminated hitherto by any foreign thinking on locomotive design. G. E. Scholes was his chief assistant, a man of the utmost integrity and objectivity with whom it was a delight to work. Finally on the Southern was the combination of C. S. Cocks with W. Durban who had both been tried in the fire of Bulleid's unconventional activities, and who had only now been released from the agonies of trying to design something workable out of the ideas behind the "Leader" locomotive.

This was the team which undertook the production design, subject only to one change early on, when on Coleman's retirement, C. S. Cocks was put in charge of design at Derby, and R. G. Jarvis, formerly Chief Technical Assistant at that centre, transferred to Brighton in his stead.

Thus production design became the responsibility of five men, the chief locomotive designers at Derby, Doncaster, Swindon and Brighton, together with myself. I do not think of it as design by committee, for each of us had a large and important segment of the work under his individual control. It was necessary for us to meet regularly at headquarters, however, in what came to be known as the Chief Draughtsman's Committee, while as already mentioned I visited each of the design offices with equal regularity. This activity was cemented by intermediate circulation of drawings of previous practice, reports on user experience, and proposals for further rounds of discussion at the next meetings. There was often spirited argument only terminated by the decision which it was my responsibility to give. The complete absence of bloody-mindedness was a pleasing feature, as this could have been a real stumbling block when considering the strong personalities and loyally-held traditions of the men concerned. Each segment of the mosaic was faithfully recorded in minutes which after approval by Riddles were made available to the Executive, to the operators and civil engineers and to the mechanical engineers in charge of the different regions for criticism or comment. Little was forthcoming however, so well had the ground been canvassed beforehand. Thus the design structure for each class was gradually built up, and in each design office, seconds in command, section leaders and draughtsmen took up their respective portions, and produced the materials lists, data sheets and working drawings which were passed into the workshops in ever increasing volume. At all stages of this procedure works management and the foremen of individual shops were consulted so that manufacturing problems were thrown up as soon as possible, and design modifications could be introduced before any quantity of the components concerned had been produced. Moreover as building, first of components, then of major items and finally, assembly of frames in the erecting shop, began, draughtsmen were in constant attendance so that difficulties could be identified and overcome immediately they appeared. While all of this had long been normal practice in locomotive building, it called in this instance for much more flexibility and initiative than usual, because drawings for any one locomotive type in a given workshop came from several drawing offices, while each drawing office had to cover building in workshops far away from, as well as adjacent to, its own physical position on the map.

That it worked out so excellently as it did is a tribute to all of the design and workshop staffs who rose to this complex occasion in such a forthcoming manner.

Having touched upon the design process and its personalities it is now proposed to describe in more detail what was designed. It would take a thick volume to cover each part of the locomotive, to describe the alternatives which presented themselves, and the discussions which led to decisions in principle, rounded off by presentation of the final design in each case. It is only possible in the space available to touch upon the more interesting features presented in relatively non-technical language for the benefit of the general reader. A little background can in most cases hardly be avoided, however, if what was decided under each heading is to be made comprehensible. In following the mental processes behind the decisions it may be possible to sense the final level attained by steam design in this country and to appreciate what were thought at that time to be the matters most needful of close attention.

Dealing first with the boilers and their associated equipment, it was decided to use four existing L.M. boilers which had already proved themselves first-class steam producers and had given minimum trouble and maintenance costs in service. These were applied as follows:—

L.M. Engine	*B.R. Engine*
44800 Class 5 4–6–0	73000 Class 5 4–6–0
43000 Class 4 2–6–0	76000 Class 4 2–6–0
42200 Class 4 2–6–4T	80000 Class 4 2–6–4T
46400 Class 2 2–6–0 }	{ 78000 Class 2 2–6–0
41200 Class 2 2–6–2T }	{ 84000 Class 2 2–6–2T

All of these boilers retained the working pressures they were designed for except only that applicable to the 2–6–4 tank. The 200 lb per sq. in. of this boiler called for cylinders $19\frac{5}{8}$-in. diameter × 26-in. stroke to produce the required tractive effort, but outside cylinders of this size, although suitable for the L.M.S., would not satisfy the new "universal" L_1 loading gauge, which would only accommodate 18-in. diameter cylinders. It was therefore necessary to raise the boiler pressure to 225 lb./sq. in., and it was found possible to do this by retaining all the existing plates, but re-staying them at a somewhat closer pitch. The L.M. and B.R. versions thus remained dimensionally interchangeable, but the latter together with all the other ex-L.M.S. boilers, were modified to take the new range of standard boiler fittings which are referred to later.

A further version of the L.M.S. 2–6–4T boiler was also evolved which was lengthened by 9 in. in the barrel whilst retaining the identical firebox to make it suitable for the Class 4 4–6–0 of the 75000 series which it will be remembered was to have the same power capacity as the tank engine but with a wider range by carrying more fuel and water in its tender.

The 43000 Class L.M. boiler was dimensionally suitable also for the Class 3 2–6–0 and 2–6–2T but it was too heavy. These engines were destined for routes limited to a 16-ton axle load, give or take a few cwt., and some 2 tons less boiler weight was necessary, particularly in the case of the tank engine which was to be built first. It was decided therefore to adapt the Swindon No. 2 boiler previously used on Western Region tank engines of the 5100, 8100 and 5600 classes. Existing flanging blocks were made use of but the barrel was shortened by $5\frac{13}{16}$ in. To save further weight the barrel rings were made from high-tensile steel plate $\frac{1}{2}$-in. thick in place of the former mild steel plate $\frac{1}{8}$-in. thick. An 18- in place of a seven-element superheater was introduced, and, in conformity with the overall policy, a dome was applied to what had hitherto been a domeless boiler. All of the firebox waterspace stays were made of monel metal, fitted with nuts inside the firebox, in place of the mixed steel and copper stays used on the original boiler. This latter change followed that made earlier on the L.M.S.R. where experience had shown that the new combination gave much improved resistance to deterioration particularly under the influence of mixed boiler feed waters.

For the remaining four B.R. designs, the Class 8, 7 and 6 4–6–2s and for the eventual 2–10–0, there were no suitable existing boilers and entirely new designs were called for having wide fireboxes. The objective was to provide grate areas which would permit average duties appropriate to each class, to be undertaken at combustion rates not exceeding 45 to 50 lb. of coal per sq. ft. of grate per hour which would assure boiler efficiencies of the order of 80 per cent with good coal. These grate areas had also to provide a good margin for acceptable performance with poorer coals. To assure plenty of steam at all times so that the engine could never beat the boiler, especially on the engines destined for fast passenger work, a high percentage of free area through the tubes to grate area was necessary—over 14 per cent if possible. In looking over the preliminary designs for the Class 6 and 7 engines, Coleman made a major contribution by suggesting a considerable increase in barrel diameter at the back end, from 5 ft. 9 in. to 6 ft. 1 in. for the smaller engine and from 6 ft. 1 in. to 6 ft. 5½ in. for the larger. The weight saved by use of only two cylinders made this increase in boiler weight possible within permitted axle loads, which allowed in turn free gas areas of 15.9 per cent and 16.2 per cent of the grate area respectively, values normally very difficult to obtain from the combination of wide firebox with combustion chamber. It was not possible to get such good proportions in the case of the Class 8 4–6–2 and the Class 9 2–10–0. Since 6 ft. 5½ in. was the largest boiler diameter obtainable over 6 ft. 2 in.-coupled wheels within the L_2 loading gauge, the barrel of the Class 7 Pacific was retained for the former, but the grate area was enlarged from 42 to 48.2 sq. ft., again for the purpose of assuring low combustion rates at average power outputs. Thus the percentage of free area through the tubes to grate area fell from 16.2 per cent to 14 per cent, still adequate, but one factor which, when added to others, produced an inferior performance for this particular boiler as will be referred to in a later chapter. On the 2–10–0 engine, the necessity of placing the grate above the trailing coupled wheels, pitched the boiler so high that diameter had to be reduced to 6 ft. 1 in. at the back end. This produced a free area ratio of only 13.6 per cent but in the case of freight working with its lower general steam demand, no detriment to the overall performance of the engine resulted.

Another apparently minor but very effective development on these new boilers was to break away from the hitherto universal diameter for large flue tubes of 5⅛ in. From the earliest days of superheating and certainly throughout the whole of the L.M. fleet, only this single diameter of flue had been used from the smallest to the largest boilers. Wagner* confirmed that there is an optimum relationship between resistance to flow of gases and heat absorption capacity when the free cross sectional area of the tube is about one four-hundredth of the total swept surface. Containing four 1⅜-in. diameter superheater element tubes, the 5⅛-in. diameter flue produces this best condition for a length between tubeplates of about 14 ft. typical of an average 4–6–0. For shorter tubes there is not enough gas resistance and boiler efficiency is lost. For longer tubes which can vary from 17 to 22 ft. in "Pacific" locomotives, resistance to the flow of gases along the large tubes is restricted, and in particular, superheat tempera-

* "Some New Developments of the Stephenson Boiler", R. P. Wagner, *Proc. Inst. Loco. E.* No. 93, 1929.

Table VI
Boiler Dimensions

Class			*71000*	*70000*	*72000*	*92000*	*73000*	*75000*	*80000*	*76000*	*77000 82000*	*78000 84000*
Barrel Dia. Outside		Maximum	6′ 5½″	6′ 5½″	6′ 1″	6′ 1″	5′ 8½″	5′ 3″	5′ 3″	5′ 3″	5′ 0½″	4′ 8″
		Minimum	5′ 9″	5′ 9″	5′ 4″	5′ 9″	4′ 11$\frac{11}{16}$″	4′ 9″	4′ 9″	4′ 9″	4′ 5″	4′ 3″
Firebox Outside		Length	8′ 0⅛″	7′ 0″	6′ 9″	7′ 5½″	9′ 2$\frac{13}{16}$″	8′ 6″	8′ 6″	7′ 6″	7′ 0″	5′ 11″
		Max. Width	7′ 9″	7′ 9″	7′ 0″	7′ 0⅛″	3′ 11⅞″	4′ 0½″	4′ 0½″	4′ 0½″	4′ 0″	4′ 0$\frac{7}{16}$″
Tubes	Small	Outside Dia.	2⅛″	2⅛″	2⅛″	2″	1⅞″	1¾″	1¾″	1⅝″	1⅝″	1⅝″
		Number	136	136	108	138	151	157	157	156	143	162
	Large	Outside Dia.	5½″	5½″	5½″	5¼″	5⅛″	5⅛″	5⅛″	5⅛″	5⅛″	5⅛″
		Number	40	40	35	35	28	21	21	24	18	12
Superheater Elements Outside Dia.			1⅜″	1⅜″	1⅜″	1⅜″	1⅜″	1⅜″	1⅜″	1⅜″	1⅜″	1⅜″
Length between Tubeplates			17′ 0″	17′ 0″	17′ 0″	15′ 3″	13′ 2⅞″	13′ 0″	12′ 3″	10′ 10½″	10′ 10½″	10′ 10½″
Heating Surface	Tubes	sq. ft.	2264	2264	1878	1836	1479	1301	1223	1075	924	924
	Firebox	sq. ft.	226	210	195	179	171	143	143	131	118	101
	Total	sq. ft.	2490	2474	2073	2015	1650	1444	1366	1206	1042	1025
	Superheater	sq. ft.	691	677	592	535	358	258	240	247	184	124
Grate Area		sq. ft.	48·6	42·0	36·0	40·2	28·7	26·7	26·7	23·0	20·35	17·5
Working Pressure		lbs./sq. in.	250	250	225	250	225	225	225	225	200	200
Firebox Volume		cu. ft.	296	273	248	213	170	131	131	115	107	83
Water Surface at ½ Glass		sq. ft.	127	122	115	115	103	87	84	77	72	63
Volume Steam above ½ Glass		cu. ft.	153	148	146	109	96	82	80	70	76	53

Table VII
Boiler Proportions

Class		*71000*	*70000*	*72000*	*92000*	*73000*	*75000*	*80000*	*76000*	*77000 82000*	*78000 84000*
Free Area through Tubes Large	sq. ft.	4·13	4·13	3·62	3·10	2·33	1·74	1·74	1·96	1·49	0·98
Small	sq. ft.	2·66	2·66	2·11	2·35	2·22	2·04	2·04	1·70	1·59	1·76
Total	sq. ft.	6·79	6·79	5·73	5·45	4·55	3·78	3·78	3·66	3·08	2·74
Area through Large Tubes as % of Total		60·9	60·9	63·2	56·9	51·1	46·0	46·0	53·6	48·4	36·5
Total Free Area as % of Grate Area		14·0	16·2	15·9	13·6	15·9	14·2	14·2	16·5	15·2	15·9
A/S Ratio Large Tubes	one over	420	420	420	447	383	368	345	334	302	343
Small ,,	one over	435	435	435	415	392	405	382	368	374	369
Steam Pipe in Boiler	Bore	7″	7″	7″	7″	6″	5½″	5½″	5½″	5½″	5″
Cross Sectional Area	sq. in.	38·5	38·5	38·5	38·5	28·3	23·8	23·8	23·8	23·8	19·6
Regulator Full Open Area	sq. in.	52·9*	48·8*	37·0*	22·2†	26·3	24·3	24·3	22·2	24·3	21·5
Superheater Elements Total Area	sq. in.	37·2	39·3	34·4	32·5	27·5	20·6	20·6	22·3	17·7	11·1
Steam Pipes to Cylinders Total Area	sq. in.	(3) 84·8	56·5	56·5	56·5	39·3	31·8	31·8	31·8	31·8	22·1
Blast Pipe Cap Diameter		(2) 4″	5⅜″	5¼″	5⅜″	5⅛″	4¾″	4¾″	4⅜″	4½″	4⅛″
Area	sq. in.	(2) 12·6	22·7	21·6	22·7	20·6	17·7	17·7	15·0	15·9	13·4
Chimney Throat Diameter		(2) 1′ 0″	1′ 3⅝″	1′ 3⅝″	1′ 3¼″	1′ 2¼″	1′ 1½″	1′ 1½″	1′ 0 11/16″	1′ 0¾″	1′ 0″
Area	sq. ft.	(2) 0·78	1·29	1·29	1·27	1·11	1·00	1·00	0·88	0·89	0·79
Throat Area as % of Grate Area		3·22	3·09	3·57	3·16	3·88	3·75	3·75	3·82	4·35	4·40
Height—Blast Pipe Cap to Chimney Throat		2′ 6⅛″	2′ 10¾″	3′ 0½″	2′ 4⅞″	2′ 9⅝″	2′ 7″	2′ 7″	2′ 7¾″	2′ 7″	2′ 5½″
Taper of Chimney		1 in 13·9	1 in 14	1 in 14	1 in 14	1 in 14	1 in 14	1 in 14	1 in 14	1 in 14	1 in 14

** Multiple Valve Regulator †As Modified*

tures tend to be lower than they ought for greatest cylinder efficiency. The manufacturers were persuaded to break away from their unvarying product, so that $5\frac{1}{2}$-in. diameter flue tubes became available for the Pacifics, and $5\frac{1}{4}$-in. flues for the 2–10–0 resulting in steam temperatures from 735 to 750°F. becoming available at maximum output in service.

The dimensions and proportions of the wide firebox boilers are shown in Tables VI and VII. These tables repay study because they indicate the compromise which has to be made between optimum proportions, and use of standard components. If for example the 1/400 proportion of area to surface of tubes were insisted upon then a bigger variety of tube-plate sizes and tube diameters would be required in conjunction with the six lengths of tube used in the series of twelve boilers. This would be very uneconomical from the first cost, stores, stocks, and maintenance points of view, so that some sacrifice of the ideal was accepted and it will be seen that A/S values range from 1/302 to 1/420 for superheater tubes and from 1/374 to 1/435 for small tubes. It is in these kinds of compromises, guided by experience and with due regard for the intended duties of the various types, that the art of locomotive design resides.

The six narrow firebox boilers based upon L.M. and G.W. practice already had copper inner fireboxes, and the same material was chosen for the four wide firebox boilers. With the improvements in design and manufacture made by Stanier upon former G.W. practice, and with the continued development under Ivatt, copper fireboxes had given excellent service on the L.M.S., particularly in the case of the "Pacifics" on that line which ran 750,000 miles and more between inner firebox renewals. Moreover although expensive in first cost, there was a high scrap value amounting to from 70 to 75 per cent of the cost of new plate for the material reclaimed at the end of the day. The steel firebox on the other hand, in universal use in America, and in wide use elsewhere had been introduced by Bulleid to his Pacific locomotives for the S.R. and these were also giving good service so long as adequate water treatment was assured. Lighter and cheaper in first cost than copper, steel fireboxes were also able, by virtue of their weldability, to avoid lapped joints, frequent source of deterioration, and to eliminate tube leakage at the firebox end even under conditions of sustained high output. It was also easy to apply circulation improving fitments such as the thermic syphon. The choice was nicely balanced, but the decision came down on the side of copper mainly because, with the indifferent and mixed qualities of water still forthcoming over the greater part of British Railways, it was clear that steel was likely to prove a source of anxiety in this particular application due to corrosion and cracking.

The new boilers having working pressures up to 250 lb./sq. in. followed closely upon the successful designs for the Stanier Pacifics and in particular very generous radii were provided at the throat joining the barrel to the firebox front. The Belpaire shape was retained at the top to assure ample volume for disengagement of steam in the area of maximum evaporation. Plate 19 shows the Class 7 boiler up-ended during building in Crewe works in a manner which displays the main features of its construction, whilst Plate 20 shows the corresponding copper inner firebox in the making, showing in particular the easy bends at the throat of the combustion chamber.

Having got a satisfactorily proportioned boiler in the middle so to speak, it depends very largely upon what is attached at the two ends for effective performance—that is upon the opportunity for entry of the required quantities of air into the firebox through ashpan and firegate, and upon the mechanism for creating a vacuum in the smokebox and so drawing this air plus the products of combustion through the boiler tubes for the production of steam. Humble component though it may be, the ashpan has proved the Achilles heel of many designs in the past because its shape and cross sectional areas were inadequate for full air flow, and its sloping surfaces when covered with ash choked the air supply to a further degree. Moreover the old process of raking out the ashes after a run by manual means was most disagreeable and was for this reason sometimes scamped. Therefore a great deal of attention was paid to the ashpan in the new designs. For the narrow fireboxes inherited from the L.M., bottom flap doors were fitted which, when turned through 90° by a portable lever applied externally, emptied the whole contents in a matter of seconds into the pit. On Pacific locomitives the rear frame extension behind the coupled wheels formed a serious obstacle both to shape and capacity of the ashpan, so a three-compartment type already developed for the Southern "Merchant Navies" was adopted. Plate 21 shows this clearly, having a central portion between the frames with an additional hopper outside each frame thus providing large ash capacity with few "dead" surfaces. The total volume of this pan was 50 cu. ft. almost twice that usual on 4–6–0 engines. For the 2–10–0 this construction was clearly impossible, and the design was conventional, the inclined slopes being disposed at as steep an angle as the disposition of firebox foundation ring, main frames and coupled wheels would allow. Although the capacity was less, the coal burned and the volume of ash produced per hour was less on freight duties than on passenger.

All locomotives were provided with rocking grates consisting of cast iron sections having air slots cast in, each 11⅜ in. long by 2¼ in. wide. These were threaded on rocker bars across the width of the firebox, and in the case of the Class 7 engine for example, six of these bars on each side of a central division covered the whole area of the grate with a porous platform on which the firebed lay. Thus of the total grate area of 42 sq. ft., 37 per cent or 15½ sq. ft. was open space through which the air passing through the ashpan was drawn in amongst the fuel. Each transverse bar assembly was coupled to its neighbours fore and aft, and to a lever in the cab whose stroke was limited by a simple pair of catches. In the short stroke position the bars could be rocked during a run by the fireman, the few degrees of movement each way which were obtained causing clinker to be broken up, and accumulated ash to be shaken out into the ashpan beneath. The full stroke position was used during disposal at the shed after a turn of duty. In this setting, the bars could be pivoted over almost at right angles, leaving large slot-like gaps through which the whole of the residual fire could fall into the ashpan, and through the bottom flap doors of the latter straight into the pit. The cut away view of the firebox in Plate 22 shows the disposition of the rocking grate with its actuating levers.

For satisfactory smokefree combustion some top air above the firebed is also necessary, varying in quantity with type of coal, firebox volume, state of the fire, etc. The means by which this was made to enter through the firehole was extraordinarily varied on the former companies, firedoors consisting of "oven doors" or sliding flaps, and operation in service on the footplate varied all the way from leaving the doors wide open all the time, to closure of the door between each separate shovel full of coal. For the standard engines the sliding type of firedoor already used on L.M. and W. Regions was retained. In the passage of time the L.M. version had been enlarged in thickness to permit twice as much air to pass through the ports when the door was shut. This was best suited to Yorkshire coals and it was applied to locomotives supplied to all regions except the Western. There, with a preponderance of Welsh coal, provision could be made for doors having a reduced air space, and this alternative was provided, the two kinds of doors with their operating levers being interchangeable.

Circular smokeboxes bolted to saddles were provided at the other end of the boilers, having considerably domed doors with centre dart as the means of fixing tight. Stanier had brought this door from the G.W. to the L.M.S., and it was also used on the Gresley and subsequent engines on the L.N.E. Each engine type was fitted with the full American self-cleaning front end. Previously without such a device, the products of combustion when they left the tubes were drawn straight up the chimney, including a proportion of incandescent cinders, and the heavier ash particles dropped to the bottom of the smokebox where they accumulated and had to be cleaned out after every day's working. The self-cleaning device (Plate 23) consisted of removable plates set across the smokebox interior in such a way that the whole products of combustion with suspended ash were drawn through a restricted crescent shaped opening in the lower part of the smokebox which increased their velocity, and then through large wire mesh spark-arresting sieves set across the smokebox before they finally mingled with the exhaust steam and escaped up the chimney. In passing through the sieves the red hot cinders were bounced against the wire mesh by the action of the draught and were cooled sufficiently to escape finally through the chimney in the form of fine ash in a black state incapable of setting fire to anything. When properly proportioned, the self-cleaning front end would keep the smokebox uncluttered by ash over the boiler washout period of seven to 14 days. Nothing is obtained for nothing in this world, and the price which was paid for reduction in the fire hazard, and elimination of daily smokebox cleaning was an increase in the amount of draught required and the sprinkling of a certain amount of fine ash over the countryside, and sometimes on the customers through open carriage windows. At the moderate rates of combustion usual in this country, this feature did not prove a very serious drawback, although it was somewhat anti-social and hardly conforming to the spirit of the Clean Air Act!

The heart of the whole steam-producing plant was the blastpipe and chimney. This beautifully simple device, without moving parts, was nevertheless very critical as to disposition and dimensions, and while finite in its proportions on any given locomotive it had to provide the right draught to equate steam production to steam demand over the whole working range which the boiler dimensions and proportions made possible. What was by no means simple was to proportion blastpipe and chimney beforehand to assure optimum results. Before the availability of test plants, this exercise was undertaken by rule of thumb in this country, and in some countries abroad by the results of mathematical calculations of the most formidable proportions. As these calculations were not always founded on well observed data, they did not invariably produce better results than did the rule of thumb, and many indeed are the locomotives which have run their whole lives at less than their potential capacity.

Fortunately the design of the standard engines coincided with the work of S. O. Ell at Swindon, who by the development of improved means of testing steam locomotives on the stationary plant and on the line under controlled conditions, was able to study the art of draughting to a penetrating degree not previously available.* G.W. engines, and especially under the influence of Stanier, L.M.S. engines also had arrived at a relatively satisfactory steaming position by ad hoc methods—that is they were rarely actually short of steam for their daily duties. Under the harrow of Ell's analysis however, a number of engine types on both systems had been found to exploit only a meagre proportion of their inbuilt capacity. The interesting point was that quite minor modifications to chimney shape and size could produce most dramatic results. For example, a reduction in the chimney diameter at its smallest point, known as the choke, from 1 ft. 2¼ in. to 1 ft. 0¾ in. and a reduction in the amount of taper of the chimney, raised the maximum continuous steam production of the Ivatt Class 4 2–6–0 engine (already itself improved by substitution of a single chimney for the original double chimney arrangement) by 89 per cent, the blastpipe dimensions and all boiler proportions remaining unchanged. More dramatically, the W.R. "Manor" class 4–6–0, when put to the question, was found unable to sustain more than 10,000 lb. of steam per hour continuously. In this case a reduction in blastpipe oriface accompanied the narrowing of the chimney proportions, and when this was done no less than 22,000 lb. of steam per hour were then forthcoming. With this kind of work in the background it was possible to specify the draughting conditions throughout the whole range of standard engines with some confidence, all making use of a single nozzle and chimney.

Prior to Ell's work, it had been proposed to fit the larger standard

* "Developments in Locomotive Testing", S. O. Ell, *Proc. Inst. Loco. E.* No. 235, 1953.

engines with double chimneys as Plate 24 portrays, but this was based upon previous L.M. experience where this arrangement had shown marginally more favourable results on the "Duchess" Pacifics and the Class 5 mixed traffics. However, Ell was able to show the way to better results with correctly proportioned single chimneys than had been obtained formerly with some of the double arrangements, and it was decided to initiate the new series with the simple single arrangement leaving to future development work any trend to multiple nozzles at a later date. Table VII sets out relevant dimensions on the locomotives as first built. Bearing in mind the extremely critical nature of these dimensions, certain adjustments to them were found necessary in some cases as the locomotives went into service and could be thoroughly examined on the test plants. Such modifications as were made will be referred to in Chapter VII.

Before passing on to choice of boiler mountings, two other important general design features may be mentioned. One was a more generous provision than hitherto on all boilers of washout plugs and doors, for clean boilers were the very essence of low maintenance costs. On the Pacifics in particular the washout doors at the rear corners of the firebox were left extremely accessible, by raising the cab footplating above them, a step which gave considerable offence to those who judged locomotive appearance by having the bottom levels of cab and tender tank more nearly lineable. Large bore manual blowdown valves, having silencers fitted to their outlets were also provided on the firebox legs, so that after each turn of duty the water spaces could be blown clear of accumulated sludge.

The other measure was to provide more sensitive regulators for the Pacifics. Judging from much available experience that locomotives of this kind tended to be slippery when rail conditions were bad, it was decided to fit the Superheater company's multiple valve regulator incorporated in the superheater header. This was standard in America and was in world wide use. A preliminary experiment on five Eastern Region V_2 class engines had confirmed that its operation was so finely graduated as to enable the driver more readily to control slipping. Because the operation of this device called for external rodding and a regulator handle in the cab working in a fore and aft direction, it was decided for the sake of a uniform cab arrangement to retain this means of operation for all the other standard engines, even although they retained the normal slide type of regulator valve in the dome.

As mentioned in Chapter II, a sub-committee of the Locomotive Standards Committee had been set to work on the standardisation of boiler mountings as early as January 1948 consisting of Messrs. R. J. Low (W.R.) in the chair, A. H. Edleston (L.M.R.), J. Allott (S.R.) and W. G. Amies (E.N.E.R.) and they had already covered a lot of ground before actual production design overtook them, and their work became merged with the main stream of drawing office work from July 1949 onwards. It would be tedious to survey all the many details classified under this heading, but having regard to the chairmanship of the main and sub-committees in this 18 months' period there was a natural trend to recommend Swindon fittings unless good cause and just impediment could be demonstrated why something else should be used instead. Swindon pressure and water gauges were accepted, the latter having a good feature in the form of a metal column which joined the top and bottom cocks behind the gauge glass proper, this maintaining the two parts of the gauge in perfect alignment, and making unit replacement easier. On the other hand, unlike other companies who fitted two gauges per engine, the G.W.R. had fitted only one, and relied upon try-cocks on the hollow vertical column above mentioned to indicate the water level in the event of failure of the watergauge as a whole. Anyone who has let the air out of their central heating radiators at home will appreciate the difficulty of judging when the air flow stops and the exit of water begins. Try-cocks as a means of judging water level in a boiler dated from the very earliest days of the locomotive, before satisfactory visual glasses became available, and with modern boiler pressures it was anybody's guess whether steam or water was emerging when the cock was opened. As a matter of fact it was exceedingly difficult to establish whether Western enginemen ever used these anachronisms at all. So the column was retained but made solid, the try-cocks were eliminated, and two standard water gauges were specified for each of the new locomotives.

Selection of live steam injectors was only made after a series of tests on the Swindon works steam fittings test plant in which both regional and proprietary instruments were tried out for range of working and for maximum delivery capacity. In this case the G.W. injector showed a marked superiority over all the others, and it was adopted in three sizes to cover the proposed fleet, giving maximum water deliveries of 34.800, 25.700 and 18.500 lb. per hour respectively, a useful range of adjustment of the quantity delivered being available in each case. It is interesting to recall that this injector design was one of the items which Stanier did not transplant to the L.M.S. when he joined it in 1932, but the Derby injector which he retained for his new command was at this late hour found to be markedly inferior.

The only mildly controversial item on this standardisation of boiler mountings was the top feed clack valve fitted at the point where the injector delivery entered the boiler. The sub-committee put forward the Swindon type which had been a standard fitting on that line since Churchward's time. This was however one of the Western specialities which did not transplant well, and, in a number of successive versions in Stanier's time and after, the L.M.S. had failed to get satisfaction for it was super sensitive to assembly by the ham-handed, and leakage of feed water inside the boiler clothing was an all too frequent occurrence. On the other hand nobody had ever seen the very simple straightforward clack valve used by the Southern Railway leak, whether applied to boiler side or top. The sub-committee was therefore over-ridden on this item, and the Southern design adopted. It can be said at this juncture that it

29

Underside of footplate of a Class 7 locomotive showing the pipework.

30

Full size wooden mock-up of the rear end of the engine and front part of the tender of the Class 7 4–6–2.

31 View inside the cab of the full size wooden mock-up of the Class 7 4–6–2.

33 Rotating jig for welding horn guides into the main frame at Crewe Works.

32 Close up view of a mechanical lubricator and two groups of atomisers with diaphragm type check valves.

35 A Class 5 4–6–0 under construction at Derby.

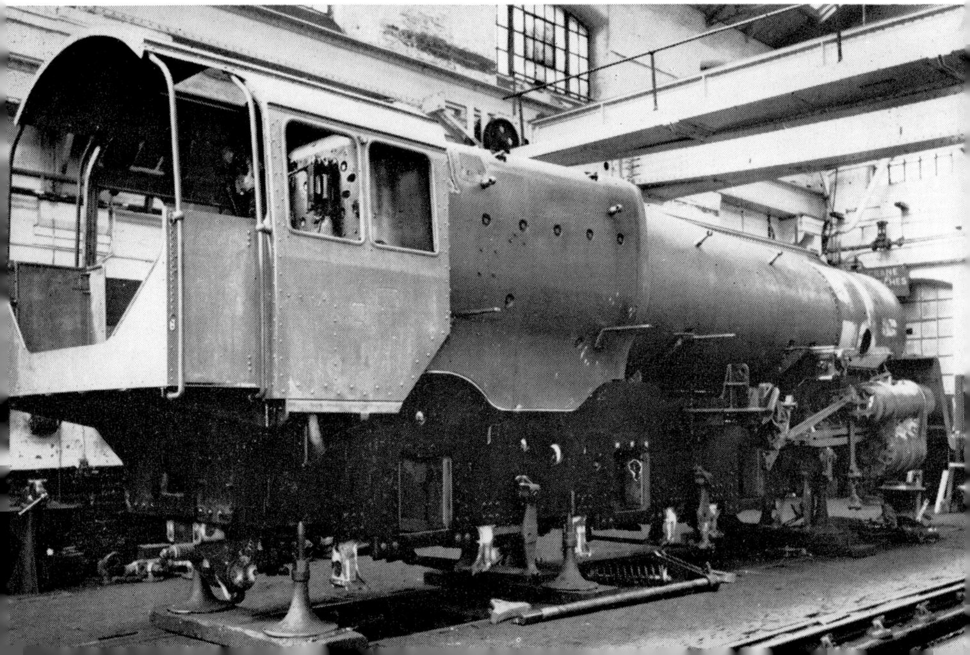

34 A Class 9 2–10–0 under construction at Crewe.

36 A Class 4 4–6–0 in Swindon erecting shop.

38 Inspection of No. 75000 new out of Swindon Works. From left: Messrs Millard (erecting shop foreman), Roberts (works manager), Riddles (Railway Executive member), author, Jarvis (chief technical assistant, Brighton), and Durban (chief locomotive draughtsman, Brighton).

37 A Class 4 2–6–4T in Brighton erecting shop.

39 The naming ceremony of No. 70000 at Marylebone Station on January 30, 1951. From the left (on platform): Sir Eustace Missenden, Lord Hurcomb, R. A. Riddles and the Rt. Hon. Alfred Barnes, Minister of Transport.

40 Inside connecting rod of 3-cylinder 4–6–2 No. 71000.

41 Class 5 4–6–0 No. 73150 fitted with Caprotti valve gear. *[P. J. Sharpe*

42 Class 7 4–6–2 No. 70044 temporarily fitted with Westinghouse air pumps.

43 Class 9 2–10–0 No. 92028 with a Crosti boiler in original condition working an up coal train near Ampthill Tunnel, south of Bedford.
[E. S. Cox

44 Arrangement of the Berkeley mechanical stoker on a Class 9 2–10–0.

45 Class 4 4–6–0 No. 75029 with the Swindon design of double chimney.

46 Class 4 4–6–0 No. 75070 with the shorter Brighton design of double chimney, near Oxted, Southern Region, with a down East Grinstead train. *[E. S. Cox*

47 Class 4 4–6–0 No. 75006 on the Swindon test plant in November 1951, during the running-up period of a test at 19,600lb of steam per hour.

48 Class 4 4–6–0 No. 75006 during a controlled road test with dynamometer car on the Western Region.

49 Class 7 4–6–2 No. 70005 on the Rugby test plant.

50 Class 8 4–6–2 No. 71000 *Duke of Gloucester* in service, pulling away from Rugby in September 1958 with the down *Midday Scot.*
[E. Williams

51 No. 71000 *Duke of Gloucester* near Didcot hauling 20 coaches on a controlled road test.

52 Class 6 4–6–2 No. 71001 *Clan Cameron* with a morning northbound freight climbing from the Clyde viaduct towards Lamington Station in November 1960. *[D. Cross*

53 Class 7 4–6–2 No. 70013 *Oliver Cromwell* arriving at Penrith with the Keswick portion of the down *Lakes Express* on August 14, 1964. *[D. Cross*

54 Class 9 2–10–0 No. 92220 *Evening Star* standing at Newport on July 1, 1960 with the down *Capitals United Express.* [J. Hodge

55 Class 9 2–10–0 No. 92024 with a Crosti boiler passing Elstree & Borehamwood with an up Midland coal train in October 1963. [B. Stephenson

never gave the slightest trouble or leakage on the standard engines whether assembled by experts or run of the mill fitters.

Exhaust injectors of the well-proven Davies and Metcalfe type were fitted to the three Pacifics and the Class 5 4–6–0. These were considered as the engines out of the whole series most likely to experience sustained running, and were thus able to achieve the economies which this equipment was able to provide under favourable circumstances.

Opportunity was taken on the Pacifics to attach the injectors to the firebox foundation ring in the manner shown in Plate 25. Since the cab and footplating were also attached to the boiler, as will be later described, all the injector pipe work from steam manifold via the injectors to top feed clacks could also be securely anchored to the boilers, thus avoiding the differential movement and vibration which could result from this assembly being partly sited on the boiler and partly on the mainframes, as was more usual practice.

Although at first sight a mixture of practices, the ten boilers which supplied the 12 standard types with steam were actually of the purest lineage, tracing their descent from Churchward through Stanier and Ivatt in a continuous line of development and improvement. Four of them already existed and served large groups of regional engines. Two more were adapted from regional designs by a small alteration in barrel length. The remaining four were new creations in the same family of design with many common features. All of them accepted the standard range of boiler fittings and mountings.

V

Design of the Mechanical Parts

ALTHOUGH THE word chassis pertains more to the automobile than to the railway engine, it may here be accepted as a convenient word to cover the whole of the locomotive other than the boiler, and its appendages. In this chapter, therefore, we continue our survey of the various design features and the reasons for their adoption.

I will deal first with the cylinders and valve gear for the 11 two-cylinder types, leaving reference to the three-cylinder Caprotti arrangement of the Class 8, 4–6–2 to the next chapter which deals with variations from the main theme of standardisation. The cylinder sizes were in most cases the largest which could be accommodated within the particular loading gauges which were applicable, and ranged from 20 in. × 28 in. down to 16 in. × 24 in. Large straight ports and passages characterised all these cylinders, freedom for steam flow being emphasised by using 11-in. diameter piston valves for cylinders between 19 in. and 20 in. diameter, and 10-in. diameter valves for the 17½ in. and 18 in. cylinders. Although these large valves had the effect of raising clearance volumes marginally above 10 per cent, any detriment from this was considered more than balanced by the easement thus given to the passage of steam. With the same objective, the volume of the steam chest was made equal to half the piston swept volume, and in certain cases the casting was swollen between the steam ports to achieve this. The inside admission piston valves carried six narrow rings on each head, an arrangement well proven in obtaining minimum internal steam leakage as wear set in. Differential sized heads were used to facilitate withdrawal, the front heads being ⅛ in. larger than the back. The box type pistons were originally all cast iron, and were assembled on the piston rods by butting up against a collar on a parallel shank, security in position being assured by right and left-hand nuts producing a secure lock. This arrangement, evolved by George Ivatt when C.M.E. of the L.M.S., made it easy to inspect or change piston rings or heads at the depots without breaking the crosshead joint. This could be done by removing the gudgeon pin and easing the rod and piston assembly forward until the latter stood out beyond the front of the cylinder casting, a facility much appreciated by shed staff. In the case of eight out of the 12 types, the three 4–6–2s, the 2–10–0, the Class 5 4–6–0, the Class 3 and 4 2–6–0s and the Class 3 2–6–2 tank the cylinder castings themselves were made of steel, having cast iron liners inserted to carry pistons and piston valves. This practice universal in later American

Table VIII

Particulars of Cylinders and Valve Gear—Two Cylinder Engines

Class		70000	72000	92000	73000	75000	80000	76000	77000 82000	78000 84000
Cylinders—Diameter × Stroke	in.	20×28	$19\frac{1}{2}$×28	20×28	19×28	18×28	18×28	$17\frac{1}{2}$×26	$17\frac{1}{2}$×26	$16\frac{1}{2}$×24
Piston Swept Volume (One Cylinder)	cu. in.	8796	8362	8796	7939	7125	7125	6254	6254	5134
Clearance Volume as % of Piston Swept Volume		10·3	10·8	10·3	11·3	10·8	10·8	11·6	11·6	*
Maximum Piston Thrust	lb.	78540	67196	78540	63794	57256	57256	54119	48106	42765
Steam Chests										
Volume between Valve Heads	cu. in.	4558	4558	4558	4056	3895	3912	3037	3037	*
Volume as % of Piston Swept Volume		51·8	54·5	51·8	51·1	54·6	54·9	48·5	48·5	*
Piston Valves Diameter		11″	11″	11″	11″	10″	10″	10″	10″	8″
Steam Lap		$1\frac{1}{16}$″	$1\frac{1}{16}$″	$1\frac{1}{16}$″	$1\frac{1}{16}$″	$1\frac{1}{16}$″	$1\frac{1}{2}$″	$1\frac{1}{2}$″	$1\frac{1}{2}$″	$1\frac{5}{16}$″
Lead		$\frac{1}{4}$″	$\frac{1}{4}$″	$\frac{1}{4}$″	$\frac{1}{4}$″	$\frac{1}{4}$″	$\frac{1}{4}$″	$\frac{1}{4}$″	$\frac{1}{4}$″	$\frac{1}{4}$″
Exhaust Clearance		Nil	Nil	Nil	Nil	Nil	Nil	Nil	Nil	Nil
Maximum Travel of Valves		7·73″	7·73″	7·87″	7·73″	7·34″	6·58″	6·25″	6·25″	5·92″
Maximum Cut-off %		78·0	78·0	78·0	78·0	75·0	75·0	75·0	75·0	78·0
Travel at 20% Cut-off		4·05″	4·05″	4·06″	4·05″	4·05″	3·66″	3·62″	3·62″	3·23″

**Former L.M.S. Design—Records not available*

engines, was unusual in this country, having previously been employed only for certain replacement cylinders for Midland Compounds, and for Ivatt's Class 4, 2–6–0s on the L.M.S. Such cylinders were almost indestructible, and bid fair to last the life of the locomotive to which they were fitted, wear being accommodated by renewing the cast iron liners. The remaining types retained cast iron cylinders as a matter of workshop manufacturing convenience.

For engines which only had to satisfy the larger L_2 loading gauge, the three-bar slidebar arrangement of the Gresley engines was adopted, this elegant design permitting a crosshead of minimum weight. On engines with pony trucks which had to pass the smaller L_1 gauge and where the rear end of the cylinders abutted on the leading coupled wheels, the need to find lateral clearance for the coupling rods to pass between wheels and slidebars made it difficult to retain the three-bar arrangement, and the Laird design having two bars, both above the cross-head, as on the W.D. 2–8–0s was at first proposed. Even this, however, did not permit very good bearing surfaces, and it was finally agreed to use the L.M.S. arrangement having one slidebar above and another below the crosshead.

The well-tried Walschaert valve gear was used throughout, which for accessibility, smallness of wear in service, and accuracy of valve events, was hard to beat. The unprecedented travel in British practice of $7\frac{3}{4}$ in. in full gear on the larger engines was used in conjunction with $1\frac{11}{16}$-in. steam lap, and a starting cut off in forward gear of 78 per cent was obtainable. It is interesting to note that at 20 per cent cut off, the valve travel was still over 4 in., a value which was hardly exceeded in full gear by countless generations of older locomotives. These generous valve events were the principal features which distinguished the B.R. from the L.M. Class 5, 4–6–0 locomotive, in all other respects so closely similar. The smaller locomotives had $1\frac{1}{2}$-in. lap and around $6\frac{1}{2}$-in. travel, amply sufficient for the lesser duties which fell to their lot. Table VIII sets out cylinder and valve gear particulars for the engines under consideration and Plate 26 illustrates the Class 7 4–6–2 arrangement as a typical example.

Connecting and coupling rods call for no special mention, being of fluted section made in fine grain steel of 50 tons/sq. in. tensile strength, but the reversing gear was a departure from previous practice, where the screw was in the cab, operating the weigh shaft by means of a long rod running alongside the boiler. It was considered that vibration and wear could be much reduced by mounting the screw forward, and acting directly upon the weigh shaft, and Plate 27 shows very clearly how this was done. The shaft running back to the cab now simply revolved and was not subject to any kick from the valve gear. There was not room, however, to do this on the tank engines, which had perforce to retain the previous arrangement.

The main frames, backbone of the whole engine, came in for the most careful consideration. Because the traditional plate frame, in universal use on British Railways, had a very mixed record in which fatigue cracks and subsequent expensive repair played far too large a part, it was at first proposed to use bar frames for all of the larger engines with wide fireboxes. In place of rolled plate from 1-in. to 1½-in. thick, the bar frame was usually a forging from 4-in. to 5-in. thick and by virtue of its much greater strength laterally, it was less prone to bend and thus to crack. Although under extreme conditions on the largest American engines bar frames had reached the limit of their endurance, there was a lot of evidence about the world that under moderate loadings they could be expected to last the life of a locomotive without trouble.

As Plate 24 shows, work was thus started on the basis of alternative forged steel or cast steel frames, the specialist manufacturers being consulted in each case. While satisfactory designs could be evolved, it was soon found that a production bottleneck would occur because available machine capacity and handling space in railway workshops was insufficient. Even if supplied new by industry, as was perfectly possible, the frames would still require handling in railway shops later on for maintenance purposes and when it became clear that both cost and weight would be increased, the idea was abandoned.

On the other hand it was decided to retain the advantage which wide fireboxes above the frames made possible, of narrowing the frame spacing to coincide with the centre lines of the axleboxes, thus avoiding offset loading through the spring hangers. Bulleid had already developed for his "Merchant Navies" a frame of this kind having horn-guides welded in position, and Plate 28 shows how his technique was applied to the B.R. Pacifics. This plate shows the very substantial cross bracing which was designed to resist the racking action of the piston thrusts, and it also shows the thicker shallower rear extension frames which, passing below the firebox, left an entirely clear space for the provision of a deep ashpan.

On the remaining locomotives with narrow fireboxes, the frames had perforce to be sandwiched between the 4 ft. 0½ in. or so outside width of the firebox and the 4 ft. 5⅝ in. which is the distance between the backs of the coupled wheel tyres. For this reason the hornguides occupied the traditional position offset towards the interior of the frames. Although much consideration was given to the possibility, the guides were not welded to the frames as in the case of the Pacifics, but were made as steel castings bolted in position—again as a means of deploying available workshop capacity rather than as a question of superior technical merit. Great care was taken to avoid past error as far as humanly possible, and besides using the greatest thickness of plate which the permissible weight would allow, the hornstays which close the frame gap below the axleboxes received particular attention both in design and manufacture to ensure, once again, that the effects of strain did not undermine the tightness of the assembly. While it is hardly correct to think of locomotive main frames as a quivering mass, nevertheless when the engine is in motion every plate, casting, rivet and bolt is subject to reversals of strain in a manner which has only partially been charted either by theory or by strain gauging, so that the placing of the different members and their connections one to the other in the course of design has been as much a matter of art as of science. It is for this reason that the all-welded frame gained little acceptance in steam practice, even although at first sight it seemed an obvious development. But weld metal is more akin to cast than to wrought material, and as such is not very ductile. In the frame layout which the traditional form of the normal steam engine made obligatory, the breathing capacity of the riveted or bolted joint gave much more margin for design error, where the strains and forces involved were still only imperfectly mapped out. This is a piece of unfinished business to which more attention would have had to be given, had steam traction continued.

The wheels and axles showed no departure from current practice on one or other of the regions. The tyres, 3 in. thick when new, were shrunk on to cast steel spoked wheel centres having triangular-shaped rims, and uniform support of the tyre all around the circumference was obtained in the case of the coupled wheels by building up the balance weights separately from plates riveted across the spokes and containing lead weights sandwiched between them. A Bulleid feature was adopted in that the tyres were innocent of all means of fastening to the wheel centres other than by the effect of the shrinkage fit itself, supplemented by shallow lips abutting on to the wheel rim. Choice of coupled wheel diameter was largely arbitary, 5 ft. 8 in. and 6 ft. 2 in. being marginally preferred in the larger sizes over the alternative 5 ft. 6 in. and 6 ft. 0 in. No wheel smaller than 5 ft. 0 in. was used throughout the series, with 5 ft. 3 in. as a fourth size introduced simply because it was proposed to standardise the L.M.S. Class 4 2–6–0 as it stood with only detail alterations. Having adopted this size for this reason, it was convenient to apply it also to the Class 3 engines. The original proposal was that all carrying wheel and axle sets whether bogie, trailing truck or tender should have 3 ft. 0 in. diameter wheels with outside roller bearings. While outside bearings have been almost universal on tenders, and trailing trucks, they have only rarely been used on engine bogies, and there has been a tradition of relatively large diameter wheels in each case so as to reduce the peripheral speed and thus proneness to overheating of plain white metalled bearings. With roller bearings, however, the smallest practicable wheel could be used. Considerable maintenance advantage was visualised in having only a single interchangeable wheel set for all carrying positions, but in the working out of the designs this promising proposal got lost, mainly due to difficulty in accommodating outside axleboxes on so small a wheel diameter within the universal L_1 loading gauge. It should be explained that whereas adjacent to the inside vertical plane of the wheels it was possible to have parts such as axleboxes coming down to within 5 in. of rail level, outside the wheels only 6 in. was permissible, and even that over a limited width. This variation, although relatively small, made all the difference in accommodating satisfactory designs of axleboxes and their associated horn guides and underkeeps. 3 ft. 0 in. diameter wheels were therefore only retained for bogies and two-wheeled trucks with inside bearings, whereas wherever outside bearings were used, on trucks or tenders, a 3 ft. 3½ in.

Table IX

Balancing Particulars—Two Cylinder Engines

Class		*70000*	*72000*	*92000*	*73000*	*75000*	*80000*	*36000*	*77000** *82000†*	*78000** *84000†*
Revolving Masses—Weight per Cylinder	lb.	1500	1500	2040	1419	1415	1396	1316	1189	1172
Reciprocating Masses—										
Total Weight per Cylinder	lb.	846	846	865	826	760	757	737	737	564
Percentage Balanced		40	40	40	50	50	40	50	50	50
Unbalanced Weight per Cylinder	lb.	508	508	519	413	380	454	369	368	282
Ratio: Unbalanced Reciprocating Weight per Cylinder to Total Weight of Locomotive	one over	414	390	374	422	400	427	363	350* 451†	391* 526†
Hammer Blow at 5 Revs. per Second										
Per Wheel	tons	2·12	2·12	0·78	2·61	2·40	1·91	2·17	2·17	1·53
Per Axle	tons	2·55	2·55	1·56	3·11	2·86	2·28	2·58	2·58	1·82
Per Rail	tons	5·50	5·50	3·38	6·95	6·49	5·10	5·93	5·93	4·29
Whole Locomotive	tons	6·60	6·60	6·77	8·28	7·73	6·08	7·05	7·05	5·11

**Tender Engine †Tank Engine*

diameter wheel was employed. Roller bearings were applied to the whole of the carrying wheels.

Table IX gives some particulars of the balancing which was applied to these locomotives, all of the revolving, and in the case of the two-cylinder engines, a percentage of the reciprocating parts being balanced, the weight required for the latter being equally divided amongst the coupled wheels. It will be seen that the percentage of the reciprocating masses balanced varies from 40 to 50 per cent and bears some relation to the total weight of the locomotive. As regards effect upon the track, the hammer blow per axle and for the whole locomotive attains only about half the limiting values laid down by the Bridge Stress Committee of 1928 in the words "all future locomotives should be so designed that at a speed of five revolutions per second, the axle hammer blow will not exceed 5 tons, and that the hammer blow of the engine as a whole shall not exceed 12½ tons". Horizontally, theory indicated that fore and aft oscillations resulting from this compromise should not exceed the working clearances, and thus become noticeable to train passengers, unless some unsuitable characteristics of the drawbar springs existed to act as a magnifier where resonant effects were present. Adjustment of drawbar spring values was in fact subsequently found to be necessary, as will be described later, but this done, the chosen percentage of reciprocating parts balanced required no modification.

Much argument attended the question as to whether the coupled axleboxes should have plain or roller bearings. The original G.W. type of steel box with pressed-in brass bearing, whitemetal lined, having a well-type underkeep with generously dimensioned oil pad, had been so developed by Stanier and his successors on the L.M.S., in conjunction with mechanical lubrication, as practically to eliminate hot boxes. With lubricated wheel boss faces and manganese steel liners on the horn-block faces, over 100,000 miles could be run by these boxes between repairs, but this seemed to be the ceiling so far as this particular arrangement was concerned. Roller bearings of suitable size and strength to cope with the heavy thrusts of steam traction had been developed by two manufacturers, and based largely upon American and Swedish experience, gave promise of much higher mileages for the radial bearing itself, apart from the flat surfaces. But such roller bearings were very expensive, several times the cost of plain bearings. The problem was how to assess in advance which was the more economic alternative—either a box cheap in first cost but requiring relatively frequent repairs, albeit of moderate unit cost, or an expensive box, only requiring repair after very long periods, but then calling for expensive renovation or even complete renewal. After much pondering, it was decided to fit the Class 7, 6 and 5 locomotives with roller bearings on all coupled wheels, all others of the standard types having the G.W./L.M.S. design of plain bearing of ample dimensions. This particular decision was, however, made with less conviction than was usually the case, so finely balanced were the pro's and con's, and as will be seen later, when plain bearings were substituted for rollers on certain later units of these classes, no superior merit was discernible for one over the other.

Between the axleboxes and the body of the locomotive lie the bearing springs, components notorious for their short life averaging on the former L.M.S. at any rate, less than two years between plate defects or other kinds of deterioration. Laminated springs had become universal on coupled wheels as locomotives increased in size, for only by the friction between the individual plates could the necessary degree of damping be applied to the suspension system, damping which was essential to avoid the building up of oscillation to undesirable magnitude under the influence of periodically applied exterior forces. In consultation with the manufacturers, a complete new range of springs was designed, absorbing as

Table X
Representative Laminated Spring Particulars

Position	*Span*	*Range of Stress tons/sq. in.*	*No. of Plates*	*Deflection per ton (in.)*	*Static Stress tons/sq. in.*	*Engines Fitted*
Bogie	3′ 6″	6·52	21	0·218	20·1	70,000
					19·7	72,000
					22·4	73,000
					21·6	75,000
			24	0·191	22·1	80,000
Coupled	4′ 0″	7·32	20	0·229	20·8	70,000
			18	0·254	20·1	72,000
					20·5	73,000
			16	0·286	20·1	75,000
					18·8	80,000
					20·1	82,000
Trailing Truck	3′ 6″	8·37	18	0·254	22·0	70,000
						72,000
Tender	3′ 6″	8·37	16	0·286	20·6	4,250 Gal.
			15	0·305	21·3	3,500 Gal.

Notes – 1. All plates 5″ x $\frac{7}{16}$″ section except coupled wheels which are 5″ x $\frac{1}{2}$″.
2. Trailing truck springs on 4–6–2 engines have reverse camber.
3. Pony trucks on 2–6–4 and 2–6–2 Tank engines have coil springs.

much experience and know-how as was currently available in the country. Leading particulars of some of these springs are set out in Table X and these indicate that the aim was to reduce the range of stress and also the maximum stress as compared with much of current practice, while somewhat increased deflections per ton were employed. A uniform plate width of 5 in. was selected in two thicknesses, $\frac{1}{2}$ in. and $\frac{7}{16}$ in. After long experience with silico manganese steel without any ascertainable advantage, plain carbon steel was selected which could be accepted either oil or water hardened according to the manufacturing and repair conditions prevailing at any given time. Ease of handling in shops and sheds when replacements became necessary was of importance, and the design of spring hanger brackets and spring links permitted wheel sets to be lowered out of the engine frames into the drop pits in a matter of minutes. Weight adjustment was effected by ringing the changes on a graduated series of thicknesses for the cotters which formed the connection at the top and bottom ends of the spring links. Thus the integrity of the springs and facility for handling them was covered so far as seemed practicable at the time, but the resultant riding quality was rather left to look after itself. This was not due to any wanton disregard of the comfort of the engine crews, but because this is how the matter had traditionally been dealt with in an era when far less was known about the principles of vehicle suspension than has become available in the last decade. Indeed comfort vertically was less of a preoccupation than safety laterally, and the lessons of the Indian Pacific Locomotive enquiry, described in my last book were applied to the B.R. standards in full measure. The bogies, where fitted, were in the De Glehn-Churchward-Stanier tradition, with coil spring lateral control. The trailing truck of the Pacifics was to the same design as on the Bulleid Pacifics, except that, because there was at the time a lot of breakage of the coil bearing springs with which these trucks were fitted, a laminated spring was substituted. The two-wheeled trucks on other engines had bar frames and followed Ivatt's design for the later L.M.S. engines. Springs for lateral control were used on trucks as well as on bogies except only in the case of the 2–6–2 tank engines, on which one of the two pony trucks were fitted with swing links so as to break up any resonance between the fore and aft control.

So far we have dealt with the principal bones of the locomotives as it were, boiler, cylinders, main frames, wheels and axles. Now we come to the flesh which clothes them, cabs, tanks, footplating, chimney and dome contours, and the whole general appearance. One may liken to the sinews and blood vessels of the human body, the pipes and rods and associated fittings which complete the corpus, and since stopping is just as important as starting or running, some reference to braking is called for.

In contrast to the aims of pre-world war one designers, our objective in the general assembly of the new engines was to have everything which could possibly require attention in service both visible and accessible. A second purpose was to have all attachments and particularly exterior footplating and pipes and rods firmly secured against the deteriorating effects of vibration. Only with these needs fully satisfied did good appearance come into the picture. Such considerations automatically defined what the new engines would look like, rather un-British to those who found beauty in level running plates, curvaceous splashers and all pipes and fittings tucked out of sight. To another school of thought, however, the American or German aspect which exploited accessibility to the utmost, was capable by suitable proportions, of suggesting power and speed and of achieving a striking harmony in the whole entity. From 1914 onwards practical considerations had edged British design progressively away from the shrouded aspect, and nearer and nearer to transatlantic lineaments, and the B.R. standard locomotives were only the logical conclusion of this growing trend. On the other hand, accepting this trend, there was no need to indulge in brutal ugliness, and stovepipe chimneys and absence of footplating in front of the cylinders or even of any external footplating at all were quite unnecessary affronts to the fitness of things. Thus external footplating was bracketed from the boilers on the new engines instead of from the main frames. It only required a few minutes standing on a stationary test plant beside a locomotive whose wheels were rotating at the equivalent of 70 m.p.h. to realise the movements and forces which were at work to loosen the bolts and rivets of footplating attached to the main frames. On the other hand, the boiler being immensely stiff both vertically and horizontally, formed an ideal and rock-like foundation for anything attached to it. The deep coping of the footplating thus secured, served as support for a variety of pipes running the length of the engine as Plate 29 shows. Invisible in side elevation, they were all accessible from underneath.

Probably the most intensely interesting part of any locomotive is the cab, and the infinite variety in the deployment of fittings therein through design history provides a fascinating subject for the student. The cab structure itself has progressed from simple weather-board or even nothing at all, to a substantial shelter having front and side windows, with the roof extended backwards to cover the fireman at his duties. The disposition of the fittings also has progressed through the years from almost studied awkwardness and a standing driving position, to ergonomically satisfying layouts in which all needed controls and gauges are grouped adjacent to the driver's seat. As already mentioned, I had done much footplate riding before production design had commenced, and had naturally studied closely the convenience or otherwise of the cab layouts on various regional engines, and discussed with the enginemen and inspectors concerned, their views and preferences. The more modern engines of the L.N.E., L.M. and Southern all followed a more or less common philosophy in what their cabs contained. Not so, however, the Western, which strove even in its latest designs to reproduce intact the Churchward features of half a century before—the enforced standing position for driving, the low and back breaking position for the firehole door and the absence of gangway doors between engine and tender. When one did for any reason sit down on the hard tip-up seats provided, a cross wind could promote a veritable gale round the posterior, while the vertical division between the closed and open parts of the cab side had apparently been nicely positioned so that one could neither lean out nor seek shelter without most unnatural contortions. Long use and the tradition of self congratulation rife on that region, had inured the staff to these features. With an experience of footplate riding around the world, I can say however, that few footplates were so intrinsically uncomfortable and to have transferred their features to our new engines would have sparked off immediate Union uproar at any rate on the other regions.

We evolved, therefore, a footplate layout not exactly following any one region, but developing as far as possible features supported by the majority of Motive Power Department and Trades Union members whom we had consulted. We tried to arrange the firebox face in as uncluttered manner as possible, with steam supplies coming from a common fountain outside the cab, and with all controls appropriate to the driver and fireman grouped on their respective sides of the cab. In order to subject these proposals to the most realistic examination and criticism by prospective users, a full size mock-up of the whole back end of the engine and front of the tender was built in Derby Works as shown in Plates 30 and 31. This mock-up was brought to Marylebone and installed in the Railway Executive headquarters on December 5th, 1949, and was thereafter inspected by the Executive itself, by Chief Officers, by regional management and officers, and by all who would have to use the engines at depot level from district superintendents to drivers and firemen, not forgetting official Union representation. Discussion naturally ranged widely, but the arrangement was generally found acceptable, and it was applied in principle throughout all of the 12 standard types. Most controversial was the disposition of the reversing wheel. Traditionally facing the driver, Riddles had suggested that there might be advantage for ease of handling, in placing it end-on as shown in Plate 30. In his advocacy he used the homely illustration of how one's wife would prefer to operate the wringing machine, a domestic device still in some use before washing machines became as widespread as they are today. Should the reversing gear be stiff it was claimed that the wheel could be pulled round spoke by spoke more readily than if in the facing attitude. As nobody had any experience with which to gainsay these claims, this feature was accepted. The type of firedoor also came in for argument, and a separate mock-up of the Southern Region's "butterfly" type power-operated door was set up, before the sliding type was retained on a majority vote. The remaining novel feature of this cab demonstration was the extension of the floor right up to the tender front by means of cantilever supports from the boiler as seen in Plate 22, thus providing a firm platform for the fireman, and eliminating the too mobile fall plate between engine and tender which was a feature of previous practice. We did not foresee at this time that so admirable an improvement would before long draw down a measure of trouble upon our unsuspecting heads.

It has already been explained how very critical internal chimney shape and dimensions are to effective steaming of the boiler. In a like manner the chimneys external contour has the profoundest influence upon the total appearance of a locomotive. One result of Ell's excellent work on draughting was that optimum chimney diameters became relatively smaller and the taper less marked than formerly, so that a truly functional chimney would form a thin and miserable crown above the smokebox. Readers of railway periodicals will no doubt recall various illustrations at about this period showing certain well-known types of engine undergoing experimental observation while fitted with stovepipe chimneys of incredibly meagre proportions. The chimneys of the new standard engines were, therefore, all made as double walled castings, the inner wall conforming to the doctrine according to the good Mr. Ell, whereas the outside wall was formed to a larger diameter more appropriate to the liniaments of the locomotive as a whole. To decide the shape of the chimney top, full-sized wooden replicas were prepared to L.M., E.N.E. and Lancashire & Yorkshire contours and were fitted successively to a new "Fairburn" 2–6–4T currently under assembly in the Derby erecting shop. No one acquainted with the author's antecedents will wonder at the choice which was finally made, and it was heart warming to receive approval from the member of the Executive for Mechanical and Electrical Engineering to place a signature upon these engines in so gratifying a manner.

In parallel with the sub-committee working on standardisation of boiler mountings mentioned in the last chapter, there was another similar body which carried out a large amount of spadework upon general fittings down to unions, bolts and rivets, before it in turn found its work merged into regular production design. These stalwarts were Messrs. W. Harland (W.R.) in the chair, H. S. Hanson (L.M.R.), G. H. Mills (S.R.) and D. D. Gray (E. & N.E.R.) who, leading appropriate teams of draughtsmen in their respective drawing offices, worked steadfastly upon the chassis details already noted, and upon others whose description will bring this chapter to a close.

Cylinder and valve lubrication has been a problem as old as the steam locomotive itself, because of the necessarily marginal nature of this function. The large surfaces of cylinder barrel and valve liners disposed horizontally and subject to the scraping action of ring-girt piston and valve heads, offered areas which it was most difficult to keep covered with oil, bearing in mind the high temperature of the steam, its changing velocity at inlet and exhaust, and the ever present possibility of ingress of hot gases sucked down the blastpipe while coasting with steam off, and of water due to priming. Use of mechanical lubricators feeding selected points through steam atomisers has been commonplace since the dawn of superheating. Following a long period of development and testing at Swindon, certain refinements were added to this system as shown in Fig. 19.

The pumps in the cylinder lubricator driven from the valve gear, supplied oil to physically separated groups of atomisers, one serving both cylinders and the other both steam chests, as shown in Plate 32. Here the oil was met by a supply of saturated steam coming from a valve in the cab, this steam mixing with the oil and atomising it so that it entered the cylinders and valve chests as a finely divided mist. The spool-like objects seen in the plate are non-return valves which prevented any of the steam passing back to the lubricators. The atomiser steam valve in the cab was actuated by a slotted link from the regulator handle in a manner universal on former G.W. engines. A large diameter pressure gauge was attached to one of the atomiser steam pipes in the cab. This was marked to show an area in red marked "NO OIL" at a pressure of 100 lb. per sq. in. and below, and a white area marked "OIL" above that pressure. This indicator was intended to do for the driver in connection with mechanical lubrication what was formerly done in saturated

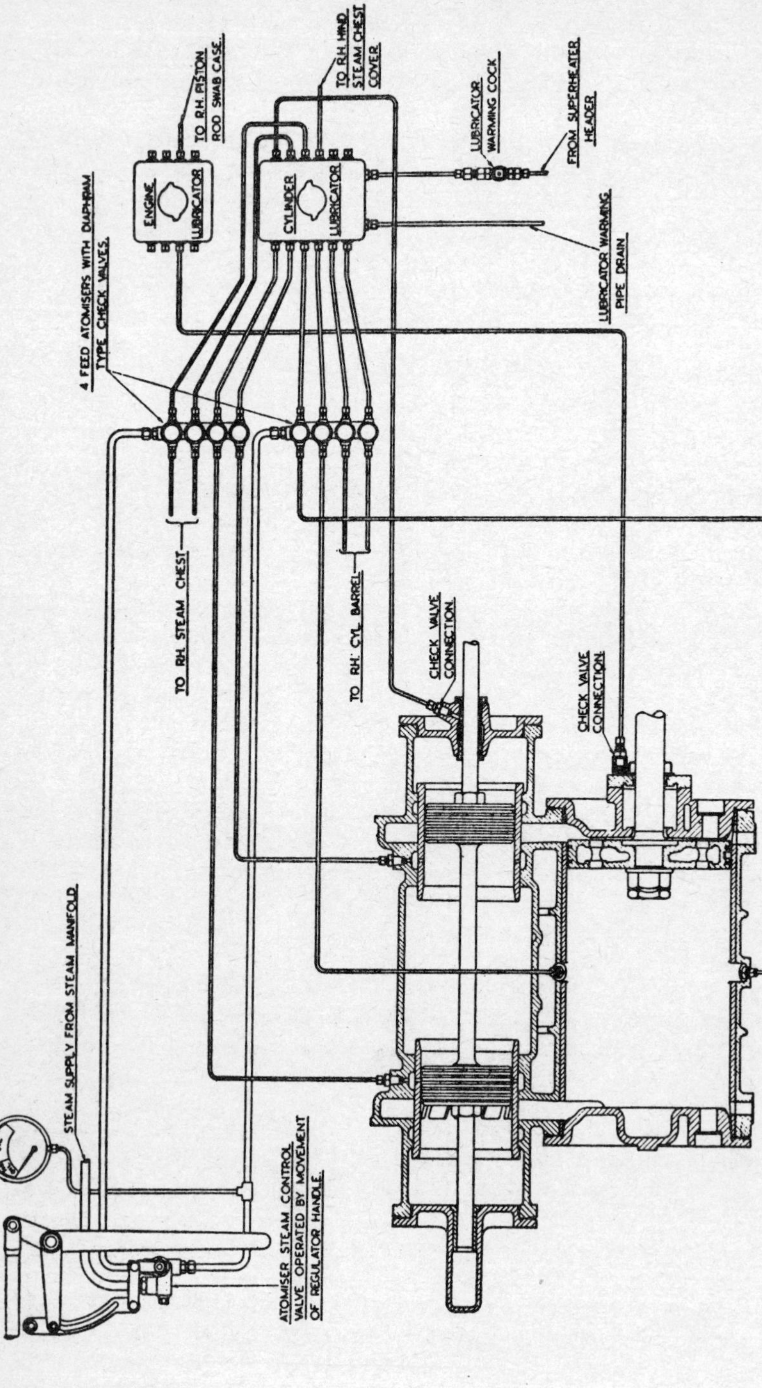

Fig. 19. Cylinder lubrication system.

steam days by the sight feeds of hydrostatic lubricators. All the time the regulator was open in any position, the atomiser steam valve was full open. When the driver closed the regulator against the stop, atomiser steam was shut off as well and "NO OIL" was shown on the gauge. This was the condition when the engine was stationary. When running with regulator closed, i.e. when coasting, drivers were instructed to move the regulator handle in such a way as to actuate the atomiser steam valve without opening the regulator itself, in which case the gauge showed "OIL", an indication which drivers were expected to maintain in all running conditions. Of course if the driver did not do this, oil still found its way to the cylinders in unatomised solid droplets, for the oil pumps worked whenever the engine was in motion, with or without steam. In this circumstance, however, not only was the arriving oil less well distributed over the surfaces to be lubricated, but the absence of the positive pressure generated in the steam chest and cylinders by the atomiser steam, encouraged entry of hot gases from the smokebox down the blastpipe with consequent carbonisation of the oil.

The question may well be asked why was all this complexity necessary, could not atomiser steam be left on all the time without devising means for shutting it off when the engine was at rest? A little thought will show however, that small though the quantity of steam was, in time it could build up in cylinder or valve chest sufficiently to cause involuntary movement of the engine, not under the driver's control, as soon as the brakes were released, a proceeding very unsafe for people working about the terminal or depot where the engine had been standing.

It can be said here and now that the regulator control for the atomisers, although a well established feature of Western practice, found little acceptance on the other regions, and to put no fine point upon it, the men would as often as not ignore the instruction and coast with the regulator handle hard shut against the stop as they had been accustomed to do with regional types. The dirty and worn valves which resulted, caused us to change the design for all the later engines, in which the atomiser steam valve was disconnected altogether from the regulator handle, and was connected instead to the cylinder cock operating handle. Thus atomiser steam was supplied in all conditions of working except only when the engine was stabled in lay-by or depot when it was customary to leave the cylinder cocks open, an action which automatically shut off the atomiser steam.

Although meticulous attention was paid to lubrication of all other locomotive parts, this only followed recent regional practice in which additional mechanical lubricators supplied oil to coupled axlebox underkeeps in the case of plain bearings. Soft grease applied by hand or power gun was widely used for the multiplicity of pin joints about the chassis. Although non-metallic and self oiling bushes were commercially available, none at that time had proved able to withstand locomotive service.

One feature which eliminated a lot of long slender rods supported by small pin joints difficult to lubricate, was the decision to operate the cylinder drain cocks from the cab by the agency of steam rather than by mechanical means. So simple and successful was this arrangement that we rather wondered why it had not been used long before.

The tenders call for no particular notice other than the fact that for the original series, the coal space was inset from the water tanks in a manner which permitted a view towards the rear from the footplate when travelling backwards. This had become commonplace on tank engines, but for tenders it had only previously been seen in any numbers in this country on the W.D. Austerity engines. When larger coal and water capacity was called for on certain duties, the tenders were later re-designed with flush sides and full-width coal spaces in the manner of former L.M.S. tenders, and it is questionable if the better rearward view given by the inset arrangement was much appreciated.

Lastly a few words on the braking system. The vacuum brake for the train was still firmly established on all non-electrified lines, and the day was as yet far off when a future British Railways Board would opt, at that late date in railway history, to substitute the air brake. The vacuum train brake had been dealt with in two different ways on the old railways so far as the locomotive was concerned, either equipping the latter with vacuum cylinders as on L.N.W., L. & Y. and G.W.R., or alternatively fitting a steam brake to the engines, which could be applied from a single control concurrently with application of the vacuum brakes on the train, a procedure favoured on the Midland (and subsequent L.M.S.) and L.N.E.R. Although the former arrangement was by far the simpler, it became increasingly difficult as engines became larger and heavier, to accommodate the large brake cylinders which the low available pressure of the vacuum system made obligatory. Certainly the combined steam and vacuum brake had been brought to a high state of effectiveness and this was adopted by British Railways. There remained a further problem in that the Western Region still operated at 25 in. of vacuum whereas the rest of the country operated at 21 in. Although much consideration was given to bringing the Western into line, it was not accomplished during the steam era, and the two vacuums were perpetuated according to allocation of the engines by simply adjusting the vacuum relief valve accordingly.

There were other differences in practice which required study and selection, as for example the combining of large and small ejectors, and the driver's brake valve into a single massive fitting mounted on the boiler faceplate as was L.N.E. and Southern practice, or to have only the driver's brake valve within the cab, and the ejectors mounted as a single fitting along the boiler side outside the cab as on the L.M.S. and Western. The separate arrangement gave much more latitude in developing a simple cab lay-out and was adopted. The different ejectors in previous use were put to the question, including a newly developed instrument from the old established firm of Gresham & Craven which they designated the "SSJ". Examples of each were put through their paces on the Swindon test plant with the result that, weighing every consideration, the new "SSJ" ejector was adopted as standard, and disposed outside the front end of the boiler on the driver's side as seen in Plate 26. On the footplate, the driver's operating handle for the combined vacuum and steam system was mounted horizontally on a pedestal convenient to the driver's right hand. Separate hand operation for the engine steam brake was retained and this in conjunction with the vacuum/steam proportional valve was the only "messy" ingredient which spoiled the neatness and symmetry of the faceplate layout as a whole.

The foregoing by no means describes all of the items which called for study and discussion on these engines, but enough has been said in this chapter and the last to give the reader a picture of the more important decisions which had to be made, and the reasons which in the circumstances of the time, led to their application.

VI

Manufacture, Total Stock and Variations from Standard

AMONGST THE vast number of locomotive books covering every possible aspect, it is remarkable how few have dealt other than incidentally with manufacture and erection. Indeed in my long connection with steam traction I can only recall one, and that written as far back as 1896. In that year George Hughes, then describing himself as an assistant in the chief mechanical engineer's department of the L. & Y.R. wrote a book* solely concerned with a blow by blow description of the building of one of Aspinall's standard 0–6–0 goods engines, from pattern and core making for the cast components up to the final erection and valve setting. In the manner of those far off days the account was dry and factual with a complete absence of interesting asides and those diversions to illuminate the general scene which more modern writers are wont to employ. None the less this book brought its author notice, and as is well known he became Chief Mechanical Engineer himself in 1904. To the best of my knowledge, nobody has attempted a similar book since then.

While the building of the B.R. series of standard locomotives could give an opportunity to make good this shortage, it would entirely unbalance this book to attempt it here, nor was this author's contact with the manufacture close enough to mark him out as the right man to undertake it. The very assiduity with which ease of manufacture was given one of the top places amongst our objectives, made certain that there would be nothing unusual to describe as regards the process of building.

All of the workshops concerned Crewe, Derby and Horwich on the L.M. region, Doncaster and Darlington on the E. & N.E., Swindon on the Western and Brighton on the Southern† had had much experience during and after the war in working upon the L.M.S. idiom, on which the design of these engines was so largely based. The home workshops as it were, could naturally cope with the new designs, and Crewe had no difficulties in the building of wide firebox boilers so closely resembling those of the Stanier "Pacifics". Workshops on the other regions had each built L.M.S. 2–8–0s during the late war, as the standard type designated by the Ministry of Supply for universal home use, while after the war, one or other of them undertook, as already described, construction of Fairburn Class 4 2–6–4 tanks, and Ivatt Class 2 and 4 2–6–0s and Class 2 2–6–2 tanks.

* *The Construction of the Modern Locomotive*, George Hughes. Published Spon, 1896.

† Eastleigh never assembled any of the B.R. standards but helped with component manufacture.

There was, therefore, very little indeed which called for anything out of the ordinary in the manner in which the components were manufactured or assembled and although the designers and builders were in almost daily contact at all stages of the work, I have no record or recollection of any major difficulty arising. Where the practice of another region had been introduced, then foremen from the building works visited their opposite numbers in the regional shops concerned in order to pick up all the know-how. The biggest novelty in this connection was the method of welding the horn guides into the main frames which had been developed for the Bulleid engines at Eastleigh, and which Crewe had now to study and apply to the 4–6–2 and 2–10–0 B.R. engines. Plate 33 shows the rotating welding jig embracing the whole side frame which was built at Crewe as a result of this co-operative effort. Wooden mockups were freely used, especially for cab layouts, and, besides that referred to in the last chapter for the Class 7 4–6–2, similar constructions were put in hand for a number of the other new types. Plates 34-37 show B.R. standard engines under construction at Crewe, Derby, Swindon, and Brighton respectively.

Of course as soon as there was anything to see at all, Riddles, Bond and myself, to say nothing of the Chief Draughtsmen concerned were quickly off to the building works, there to examine and verify, and so, from the first castings and flanging blocks through all the process of assembly, to give closest personal attention to the emerging entity. During this period the motive power people were increasingly brought into the picture, and every effort was made to have well illustrated instruction books assembled for their study before each of the engines concerned actually took the road. I recall particularly, as soon as the first draft of the book for the Class 7 engines was available, taking L. P. Parker, the motive power chief of the Great Eastern Section, which was to receive the first of these engines, up to Crewe in order to go over the whole emerging locomotive with a fine tooth comb. Parker was our first customer for these engines, and turned out to be our staunch supporter in all the vicissitudes which the future had in store for us.

There came at last the day when the first of each successive class was finished and steamed outside the erecting shop. I can think of no personal thrill so great as to see, to clamber about, and to accompany up and down the yard under its own steam, a new engine in whose inception one has been closely involved, and whose design one has controlled through the gradual build-up towards finality. To the engineer that first inspection is the red letter day, and Plate 38 shows a typical inspection, in this case of the first Class 4 4–6–0 at Swindon, where, besides Riddles and myself in the centre, are to be seen on the right of the picture the men in charge of the Brighton office who designed it, and on the left the manager and erecting shop foreman of the works which built it.

Later on management had its day and where a naming ceremony could be arranged to provide opportunity for examination by the Ministry of Transport, the Commission, and the Executive, by regional General Managers and their principal officers to say nothing of the press, then a

Table XI

British Railways—Standard Steam Engines Built 1951–1960 inclusive

	1951	*1952*	*1963*	*1954*	*1955*	*1956*	*1957*	*1958*	*1959*	*1960*	*Total*
4–6–2 Class 8 71000	—	—	—	1	—	—	—	—	—	—	1
4–6–2 Class 7 70000	25	13	7	10	—	—	—	—	—	—	55
4–6–2 Class 6 72000	2	8	—	—	—	—	—	—	—	—	10
4–6–0 Class 5 73000	29	1	20	25	44	30	23	—	—	—	172
4–6–0 Class 4 75000	16	4	25	5	13	5	12	—	—	—	80
2–6–0 Class 4 76000	—	10	25	10	15	19	36	—	—	—	115
2–6–0 Class 3 77000	—	—	—	20	—	—	—	—	—	—	20
2–6–0 Class 2 78000	—	4	8	33	10	10	—	—	—	—	65
2–6–4T Class 4 80000	17	37	18	36	23	20	4	—	—	—	155
2–6–2T Class 3 82000	—	20	—	12	13	—	—	—	—	—	45
2–6–2T Class 2 84000	—	—	20	—	—	—	10	—	—	—	30
2–10–0 Class 9 92000	—	—	—	32	38	45	56	62	15	3	251
Total	89	97	123	184	156	129	141	62	15	3	999

scene was enacted, such as Plate 39 portrays, in which the reigning Minister of Transport—in this case the Rt. Hon. Alfred Barnes—was doing the honours. Following all this, it was down to work by all concerned and out into service steamed the new machines—like their countless predecessors as perfect as their designers and builders knew how to make them, and, like their predecessors also, destined to face the hard facts of daily service and presently to display those aspects in which we had not wrought so perfectly after all! How they performed, what went well and what badly in their behaviour will be found in the following chapters. In the meantime, however, this is a convenient point at which to tabulate their running and total numbers, and to describe such major variations from standard as were introduced into successive batches of these engines as built. At this stage detail alterations as a result of experience in service will not be included, for these will be more understandably referred to when we come to deal with the teething and other troubles which were experienced in service.

Nine hundred and ninety-nine British Railways standard locomotives were built in all, from 1951 to 1960, and Table XI indicates the numbers of each type which were built in each of these years. Tables XII, XIII and XIV give the annual building programmes with the running numbers of each batch, with building works, and original allocations for maintenance and operating respectively. Table XV gives the names originally bestowed upon the Class 7 and Class 6 Pacifics. These tables are largely self explanatory but a few comments may be helpful. The building programmes each year were connected with financial authorities, and the various batches concerned and their allocations for operating were the result of discussion between regional and headquarters officers as regards traffic requirements, replacement of engines due for breaking up in a given year, and so on. The allocations for building and for subsequent maintenance were the responsbility of mechanical engineering headquarters. Building works were chosen partly on suitability, i.e. Crewe, Swindon and Doncaster were equipped to handle the largest engines more readily than Derby, Darlington, Horwich or Brighton, partly also upon completion of previous commitments, when it is recalled that these workshops were also engaged on building of regional steam types, which, although diminishing in numbers, did not run out completely until 1955. There was also a steady building of 350 h.p. diesel shunters at Derby, Crewe, Horwich, Darlington and Doncaster concurrently with the above, and in 1958 there began the building of main-line diesel electric locomotives in railway shops, so soon to develop into a flood. It was these other commitments which caused the seemingly disconnected building allocations of the standard steam engines which a study of these tables will disclose.

In a similar way allocations for maintenance were made on a basis of available capacity and sometimes of propinquity, but considerable changes were made subsequently in the course of rationalisation of overall workshop resources. It will be seen that the 1952 programme included 15

Table XII

Annual Building Programmes 1951–1954 incl.—Tender Engines

Renewal Programme	*1951*				*1952*				*1953*				*1954*			
	Nos.	*Bldg.*	*Maint.*	*Op.*	*Nos.*	*Bldg.*	*Maint.*	*Op.*	*Nos.*	*Bldg.*	*Maint.*	*Op.*	*Nos.*	*Bldg.*	*Maint.*	*Op.*
4–6.2 Cl.7	70000–70014	C	LM	E	70025–70029	C	LM	W	70045–70049	C	LM	LM				
	70015–70024	C	LM	W	70030–70034	C	LM	LM	70050–70054	C	LM	SC				
					70035–70044											
4–6.2 Cl.6	72000–72009	C	LM	SC	72010–72014*	C	LM	S								
					72015–72024*	C	LM	SC								
4–6–0 Cl.5	73000–73004	D	LM	LM	73030–73039	D	SC	SC	73050–73052	D	S	S	73075–73079	D	SC	SC
	73005–73009	D	LM	SC	73040–73049	D	LM	LM	73053–73054	D	LM	LM	73080–73089	D	S	S
	73010–73029	D	LM	LM					73055–73064	D	SC	SC				
									73065–73074	D	LM	LM				
4–6–0 Cl.4	75000–75009	SW	W	W	75020–75029	SW	W	W	75050–75064	SW	LM	LM	75080–75089*	SW	ENE	E
	75010–75019	SW	W	LM	75030–75049	SW	LM	LM	75065–75079	SW	W	S				
2–10–0 Cl.9									92000–92007	C	W	W	92030–92044	C	ENE	E
									92008–92009	C	LM	LM	92045–92059	C	LM	LM
									92010–92014	C	ENE	E	92060–92066	C	ENE	NE
									92015–92029	C	LM	LM	92067–92076	C	ENE	E
													92077–92086	C	LM	LM
													92087–92096	C	ENE	E
2–6–0 Cl.4					76000–76004	H	ENE	SC	76025–76029	DO	S	S	76045–76052	DO	ENE	NE
					76005–76019	H	S	S	76030–76044	D	ENE	E	76053–76069	DO	S	S
					76020–76024	DO	ENE	NE	76070–76074	D	ENE	SC				
2–6–0 Cl.3									77000–77004	SW	ENE	NE				
									77010–77014							
									77005–77009	SW	ENE	SC				
									77015–77019							
2–6–0 Cl.2					78000–78009	DA	W	W	78010–78019	DA	ENE	NE	78045–78054	DA	SC	SC
									78020–78044	DA	LM	LM				

Notes: *Subsequently cancelled *Bldg.*—Building Works *Maint.*—Maintaining Region *Op.*—Operating Region.
C—Crewe SW—Swindon DO—Doncaster D—Derby H—Horwick DA—Darlington

Annual Building Programmes—1955-1958 incl.—Tender Engines

Renewal Programme	1955				1956				1957				1958			
	Nos.	Bldg.	Maint.	Op.	Nos.	Bldg.	Maint.	Op.	Nos.	Bldg.	Maint.	Op.	Nos.	Bldg.	Maint.	Op.
4–6–0 Cl. 5	73090–73099	D	LM	LM	73125–73134	D	W	W								
	73100–73109	DO	SC	SC	73135–73144	D	LM	LM								
	73110–73119	DO	S	S	73145–73154	D	SC	SC								
	73120–73124	DO	SC	SC	73155–73159	DO	ENE	E								
					73160–73171	DO	ENE	NE								
2–10–0 Cl.9					92097–92099	C	ENE	NE	92203–92220	SW	W	W	92221–92250	SW	W	W
					92100–92139	C	LM	LM								
					92140–92149	C	ENE	E								
					92150–92167	C	LM	LM								
					92168–92177	C	ENE	E								
					92178–92202	SW	ENE	E								
2–6–0 Cl.4					76075–76089	H	LM	LM								
					76090–76099	H	SC	SC								
					76100–76114	DO	SC	SC								
2–6–0 Cl.2					78055–78064	DA	LM	LM								

Notes: See Table XII

Table XIV
Annual Building Programmes—1951–1958 inclusive—Tank Engines

Programme	1951				1952				1953				1954			
	Nos.	Bldg.	Maint.	Op.	Nos.	Bldg.	Maint.	Op.	Nos.	Bldg.	Maint.	Op.	Nos.	Bldg.	Maint.	Op.
2–6–4 Cl.4	80000–80009	D	SC	SC	80054–80058	D	SC	SC	80069–80080	B	ENE	E	80116–80120	B	ENE	NE
	80010–80019	B	S	S	80059–80068	B	LM	LM	80081–80095	B	LM	LM	80121–80130	B	SC	SC
	80020–80030	B	SC	SC					80096–80105	B	ENE	E				
	80031–80033	B	SC	NE					80106–80115	DO	SC	SC				
	80034–80053	B	LM	LM												
2–6.2 Cl.3	82000–82009	SW	W	W	82020–82029	SW	W	S	82035–82044	SW	W	W	82045–82054*	SW	W	W
	82010–82019	SW	W	S	82030–82034	SW	W	W					82055–82062*	SW	ENE	NE
2–6.2 Cl.2					84000–84019	C	LM	LM	84020–84029	DA	LM	S				
	1955				1956				1957				1958			
2–6–4 Cl.4					80131–80144	B	ENE	E								
					80145–80154	B	S	S								

Notes: See Table XII

Table XV
List of Names Allocated as Built

Number	*Name*	*Number*	*Name*
70000	Britannia	70035	Rudyard Kipling
70001	Lord Hurcomb	70036	Boadicea
70002	Geoffrey Chaucer	70037	Hereward The Wake
70003	John Bunyan	70038	Robin Hood
70004	William Shakespeare	70039	Sir Christopher Wren
70005	John Milton	70040	Clive of India
70006	Robert Burns	70041	Sir John Moore
70007	Coeur-de-Lion	70042	Lord Roberts
70008	Black Prince	70043	Lord Kitchener
70009	Alfred The Great	70044	Earl Haig
70010	Owen Glendower	70045	Lord Rowallan
70011	Hotspur	70046	Anzac
70012	John of Gaunt	70047	—
70013	Oliver Cromwell	70048	The Territorial Army 1908-58
70014	Iron Duke	70049	Solway Firth
70015	Apollo	70050	Firth of Clyde
70016	Ariel	70051	Firth of Forth
70017	Arrow	70052	Firth of Tay
70018	Flying Dutchman	70053	Moray Firth
70019	Lightning	70054	Dornoch Firth
70020	Mercury		
70021	Morning Star		
70022	Tornado	71000	Duke of Gloucester
70023	Venus		
70024	Vulcan		
70025	Western Star	72000	Clan Buchanan
70026	Polar Star	72001	Clan Cameron
70027	Rising Star	72002	Clan Campbell
70028	Royal Star	72003	Clan Fraser
70029	Shooting Star	72004	Clan MacDonald
70030	William Wordsworth	72005	Clan MacGregor
70031	Byron	72006	Clan MacKenzie
70032	Tennyson	72007	Clan MacKintosh
70033	Charles Dickens	72008	Clan MacLeod
70034	Thomas Hardy	72009	Clan Stewart

further "Clan" class 4–6–2s, five for the Southern and ten more for the Scottish region which were cancelled before building began. It will be recalled how closely balanced were the capacities of the Class 5 and 6 engines, and until more experience was gained of what advantages, if any, the latter could give, it was decided not to extend their use for the time being. There was also a batch of ten Class 4 4–6–0s authorised and later cancelled for the Eastern Region in 1954—the first of these engines they would have had. By now, however, the shadow of "modernisation" with the big change over to diesel power was already in the offing, and contemplated extension of railcars in particular was invading the kind of services for which this kind of locomotive would be mainly used. The same comments apply to the last 15 Class 3 2–6–2 tank engines.

While the lone Class 8 engine No. 71000 appears in Table XI as having been built in 1954, it was not included in any regular building programme. It was the subject of a special financial authorisation because it was considered as a replacement of the "Stanier" 4–6–2 No. 46202 *Princess Anne* itself only recently rebuilt from the famous *Turbomotive*. As mentioned in Chapter III No. 46202 had been totally destroyed in the terrible accident at Harrow in 1952; and rather than replace her like for like, Riddles persuaded the Executive to agree that it might take the form of a prototype Class 8 engine within the new standard range of steam motive power.

Table XVI
Departures from Standard

Engine Numbers	*Nature of Variation*
71000	3-cylinders. Caprotti Valve Gear. Flush sided Tender 10 Tons Coal 4325 gals. water Coal Pusher
70035–70039	Roller bearings Driving wheels only, Plain boxes on leading and Trailing Coupled Wheels
70040–70049	Plain boxes on all coupled wheels
70024–70029	7 Tons coal 5000 gals water Inset tender
70045–70054	9 Tons coal 4725 gals. water Flush tender Coal Pusher
70043–70044	Air compressors fitted and wing plates removed. Temporary arrangement for brake tests with mineral trains
70004	Special finish for Exhibition at Festival of Britain, 1952
73125–73154	Caprotti Valve Gear
73050–73052	7 Tons coal 5000 gals. water Inset Tender
73065–73079	
73090–73099	9 Tons coal 4725 gals. water Flush Tender
73135–73145	
73080–73089	
73100–73109	7 Tons coal 4725 gals. water Flush Tender
73120–73134	
73145–73171	
73110–73119	7 Tons coal 5625 gals. water Flush Tender
73030–73031	Air Compressors fitted. Temporary arrangement for brake tests
75050–75079	7 Tons coal 4725 gals. water Flush Tender
76053–76066	7 Tons coal 4725 gals. water Flush Tender
92020–92029	Crosti Boiler
92165–92167	Mechanical Stoker
92250	Giesl Blastpipe and Chimney
92000/1/6/79	
92165/6/7/78	Double Chimney
92183–92249	
92010–92014	
92020–92029	
92030–92044	7 Tons Coal 5625 gals. water Flush Tender
92067–92076	
92140–92149	
92168–92202	
92015–92019	
92045–92059	
92077–92086	9 Tons coal 4725 gals. water Flush Tender
92100–92139	
92150–92164	

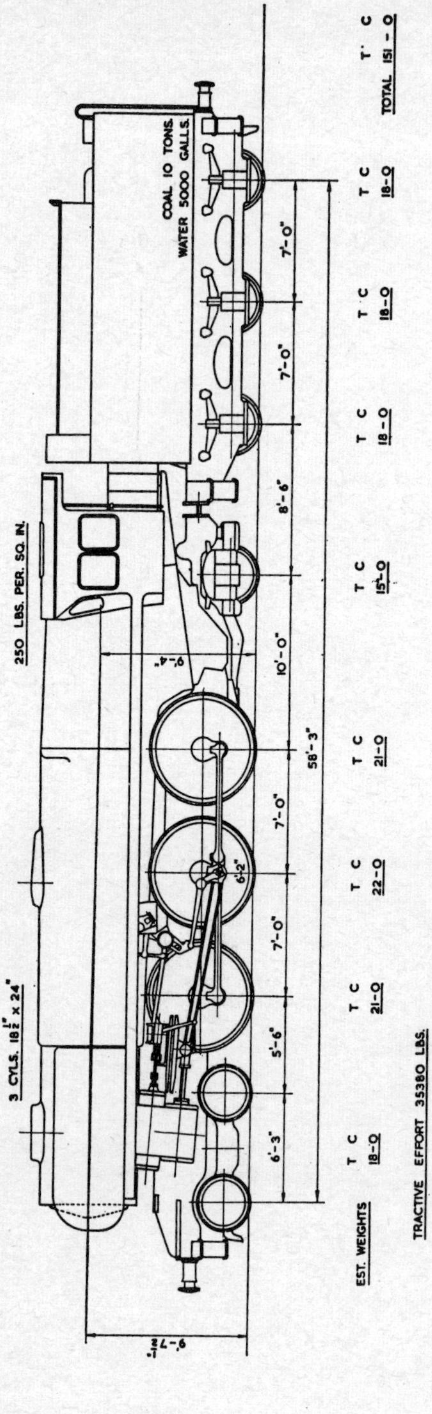

PROPOSED 3-CYLINDER 4-6-2 PASSENGER ENGINE.

Fig. 20.

This engine represented a departure from the simple two-cylinder conception with piston valves and Walschaert gear which characterised the basic standard series, and it is now proposed to describe it together with other deviations from the normal which are listed on Table XVI.

By the end of 1952 Derby drawing office had completed design work on modifying the Ivatt Class 2 2–6–0 and 2–6–2T to bring them into line with the B.R. series as regards details, and it was available to embark upon design of the big engine. By this time Ivatt had retired, and J. F. Harrison from Doncaster had succeeded him as head of the mechanical and electrical department on the L.M. Region with headquarters at Derby. In charge of the drawing offices at that centre was C. S. Cocks, whose participation in work on the B.R. engines had not extinguished recollections of Southern practice under his former master Bulleid. It came about therefore that while Riddles and I at headquarters were visualising the Class 8 as an extension of "Britannia" using as many common features as possible, the above mentioned personalities at Derby also took a hand in the game and worked out for Riddles alternative consideration the layout shown in Fig. 20 which had some strongly marked Doncaster and Eastleigh features. Harrison had been one of Gresley's young men, and had maintained throughout his career an admiration and indeed an affection for his master's products, particularly the A4 which was by common consent one of the really great locomotives of the steam era.

In true Gresley style it was proposed that the inside cylinder should drive on to the intermediate coupled wheels, and room was found within the L_2 loading gauge for the steeply inclined inside cylinder by applying a version of the "Merchant Navy" boiler, having the taper at the bottom rather than at the top of the barrel, thus raising the centre line of the smokebox to as much as 9 ft. 7½ in. There is no trace now left in the Derby drawing office of the inside valve gear which was intended, but the arrangement of the front end on the diagram could have provided equally well either for the Gresley conjugated gear lying across the front of the steam chests as on the A4s, or for a normal inside Walschaert gear similar to that which was already being thought about in connection with the proposed rebuilding of Bulleid's Pacifics. The whole confection was dressed up with Bundesbahn type smoke deflectors, a double chimney of very Thompsonian aspect, and a rather horrid looking void in front of the cylinders unmitigated by any fall plate. This apparition duly arrived at Marylebone and caused some fluttering in the dovecotes there. The reasoning behind the different features was well founded, and no doubt a successful locomotive of this kind could have been produced, but not one however which was likely to have escaped the combustion troubles at high output which eventually beset No. 71000, for reasons which will become clear as this narrative proceeds.

However, as was rather to be expected the Marylebone version of the Class 8 prevailed and the problem of the inside valve gear was overcome by driving the inside cylinder on to the leading axle and by applying

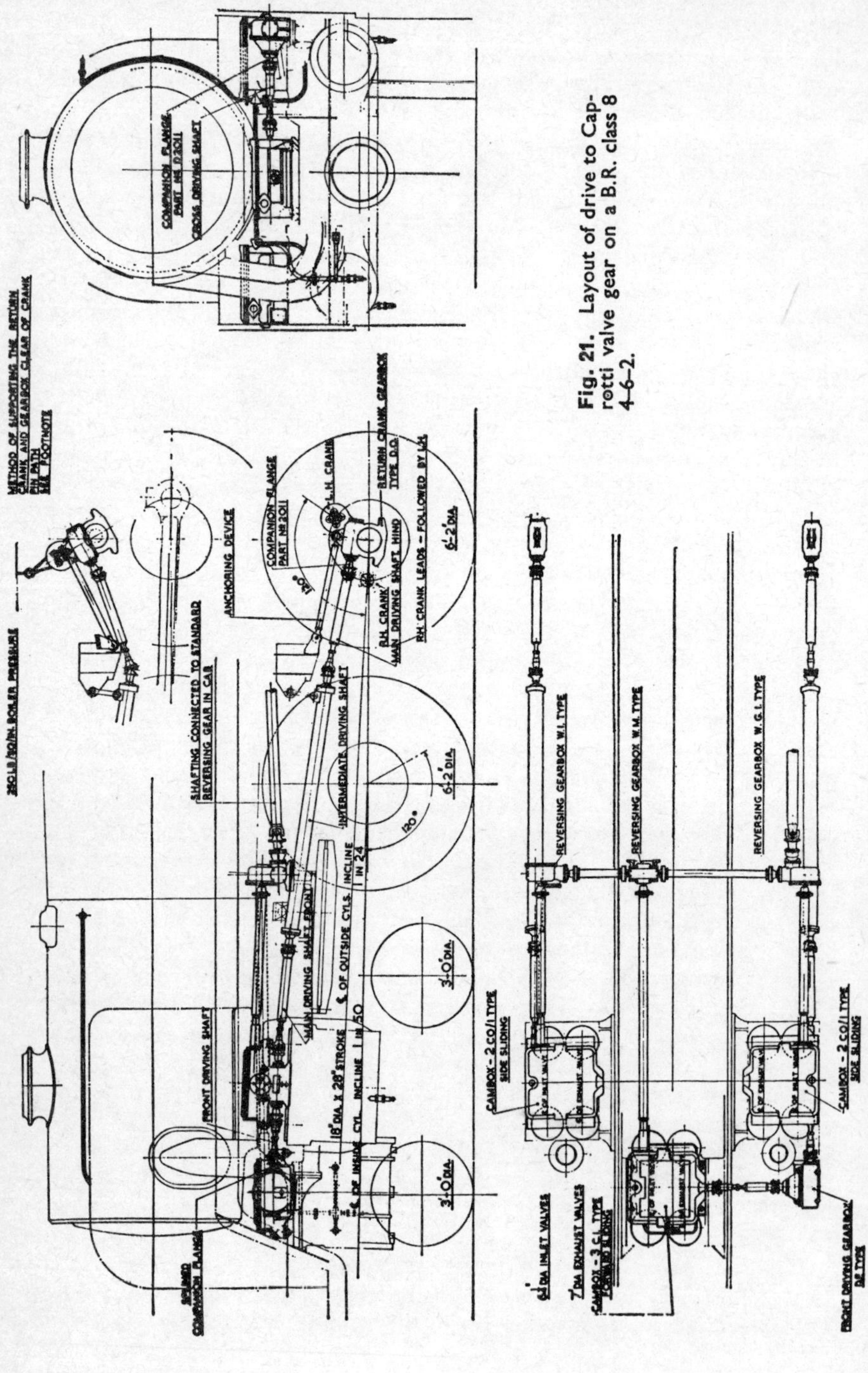

Fig. 21. Layout of drive to Caprotti valve gear on a B.R. class 8 4-6-2.

rotary cam poppet valve gear. This solved all the logistic problems of steam distribution to the the inside cylinder, and it was possible to retain the "Britannia" boiler barrel and firebox flanging blocks, an important economic consideration when only a single engine was to be built, at any rate in the first place.

The three cylinders were steel castings, the inside casting also embracing integrally the smokebox saddle and the inside slidebar carriers. A departure was made in the piston heads and rods in that the latter were flash-butt welded to the former following Gresley practice. The British Caprotti valve gear embodied separate inlet and exhaust poppet valves at each end of each cylinder, which were activated by rotary cams and very great care was taken in the design of cylinder ports and passages. The drive for the cam boxes on the outside cylinders was taken from the intermediate coupled wheels by worm-gear boxes mounted on return cranks. The drive for the inside cylinder cam box was taken from an extension on the wormshaft on the left-hand cam box through a right angle bevel gearbox. The inside cam box was placed accessibly in front of the smokebox saddle and the whole layout was as indicated in Fig. 21.

There was one other novel feature, namely the inside big end. The split brass arrangement which assembly on to a crank axle made essential had always been a troublesome feature throughout locomotive history, and the best way to avoid such trouble was to avoid inside big ends altogether. However, where their use could not be avoided as in the present case, experience had indicated that absolute rigidity of the split bearing itself was necessary and that all the means of attachment and security should be subject to a very minimum of elastic deformation and resistant to the effects of vibration. Harrison had long been interested in this subject, and it was under his personal supervision that the fork type big end shown in Plate 40 was worked out. The exceptionally robust clip closing the jaws of the rod will be noted as well as the serrated locking device for the cotter which permitted exactly the correct degree of tightness to be retained and locked up against involuntary movement. To the best of my knowledge this arrangement never gave any trouble.

In all other respects, except the enlargement of the grate area already mentioned, and the use of a double blastpipe and chimney, No. 71000 followed "Britannia" practice including 6 ft. 2 in. diameter coupled wheels, and use of roller bearings on all wheels. The tender was of the flush-sided alternative shortly to be described, and was fitted with a coal pusher. Leading dimensions of this engine have already been displayed in Fig. 11. The question of building more of these engines for the most important main-line services was raised in 1956 when there were opposing schools of thought at Headquarters as to whether there should be a pause in the acquirement of diesels for these services until more experience was gained or whether to press on regardless. Unfortunately for steam the latter policy prevailed and the *Duke of Gloucester* remained the only one of its kind.

So economical in steam consumption did this engine prove, that it was decided to extend use of this kind of steam distribution and, to gain more experience, 30 of the 47 Class 5 4–6–0s on the 1956 building programme were designated to be so fitted. Unfortunately there were some in authority at headquarters, although not in the C.M.E. department, who were determined that there should be no more development with steam, and these were last of the Class 5s to be built, the last indeed of the whole B.R. series except for the rump of Class 9 2–10–0s authorised in 1957 and 1958. The 30 "Caprotti" 4–6–0s began to emerge at the beginning of 1957, and except for retaining two cylinders, the whole of the valves and valve gear arrangements were made by the same firm and followed the same principles of design as in the case of No. 71000. Although poppet valve gears had been applied to small groups of L.M.S. and L.N.E.R. engines as far back as the 1920s, and had even been tried on the G.W.R. they had never quite rung the bell of sufficient reliability to commend them as a substitute for the conventional Walschaert or Stephenson gears. As a result of this experience, however, and of refinements introduced by their manufacturers over the years, the ultimate British Caprotti system was a thoroughly workmanlike job, and its extended use would have been practically certain had steam continued. Plate 41 shows one of these engines.

Fifteen of the "Britannias" were built non-standard so far as their coupled axleboxes were concerned. It has already been described how dubious we all were regarding the advantages of roller bearings at this position and to clinch the matter Riddles did the same thing as Ivatt had done before him with the L.M.S. Class 5s, namely to turn out five engines Nos. 70035-70039 with plain bearings for the leading and trailing coupled wheels, retaining the roller boxes for the drivers only. As a further control he had ten more engines 70040-70049 made with plain bearings on all coupled axles. Naturally this move was not very popular with the roller bearing manufacturers. The plain bearings which were fitted embodied every improvement which Stanier and Ivatt had developed, and so robust and wear-free were they, that no record or opinion I have been able to discover from motive power depots or workshops shows the slightest advantage of roller bearings over either of the other two combinations which were tried. This circumstance is perhaps the ultimate tribute to the many engineers who worked so long and persistently to win plain bearing reliability on steam locomotives.

Thinking on these lines produced the decision to use plain bearings also on the coupled wheels of the Class 9 2–10–0s authorised a year later than the above. While the lower running speeds of freight working certainly offered added assurance that all would be well, the extraordinary period in the lives of these engines in which passenger working at speeds up to 90 m.p.h. with 5 ft. 0 in. diameter wheels occurred, did not produce any distress so far as their axleboxes were concerned.

A further variation of the "Britannias" and Class 5 4–6–0s, albeit a temporary one, was the fitting of Nos. 70043 and 70044 and 73030 and 73031 with air compressors, Westinghouse drivers brake valves and the necessary auxiliary equipment for the working of brake trials with long mineral trains. It is outside the scope of this story to retail all the circumstances which led to these full-scale trials in which the air and vacuum brakes were after a long interval of years again pitted one against the other, but finding room for the double pumps which were required did not prove easy, and Plate 42 displays the rather unsightly solution which in the case of 70043 and 70044 was realised only by sacrificing the smoke deflectors.

Another unusual exception, although it can hardly be called a true variation, was the bestowal upon No. 70004 *William Shakespeare* of a super-finish (Plate 7) for its exhibition at the "Festival of Britain" which took place in 1952. For some months in that year, this engine was examined by thousands of visitors as it stood high up above the Thames on the South Bank site, and it made a brave display alongside the not very beautiful electric and diesel locomotives which were its companions. By the time, somewhat later, when No. 70004 was allocated to Stewart's Lane Shed on the Southern Region, the engine had lost something of its pristine sparkle, and it took for a while, six men six hours a day to restore it to the immaculate condition which many readers will remember when it was rostered to work the "Golden Arrow" continental service out of Victoria.

A small number of 2–10–0s were also fitted with an air compressor for the purpose of working the power-operated doors on some special bogie iron ore wagons worked in block train loads up to the Consett blast furnaces in the north of England. This, however, cannot properly be included as a variation affecting the locomotive operation itself.

Apart from the above a number of more important variations from standard were concentrated upon the 2–10–0s. The most radical of these was the fitting of Nos. 92020-92029 on the 1953 building programme with the Franco-Crosti boiler. This system boldly sought increased efficiency by modification to what was already the most efficient part of any steam locomotive, namely the boiler. The normal chimney on the front of the smokebox was closed, and the hot gases on leaving the tube bank in the boiler proper, were turned back on themselves in the smokebox, and were passed through a further series of tubes in a secondary drum before being ejected from the final chimney placed alongside the main boiler at the cab end. This secondary drum was used as a pre-heater for feed water supplied by the injectors, so that water entered the boiler proper at a temperature only slightly less than that of the water already in the boiler. This made use of a large amount of low-grade heat which would otherwise have been wasted. The cold feed from the tender was forced by the injector into the back end of the pre-heater drum, and under natural circulation the water moved up towards the front, or hot end whence it passed by way of external pipes to the main boiler clack valves. The main boiler was of normal type with no special features except that for a given output it could be made smaller than the boiler of the corresponding orthodox locomotive.

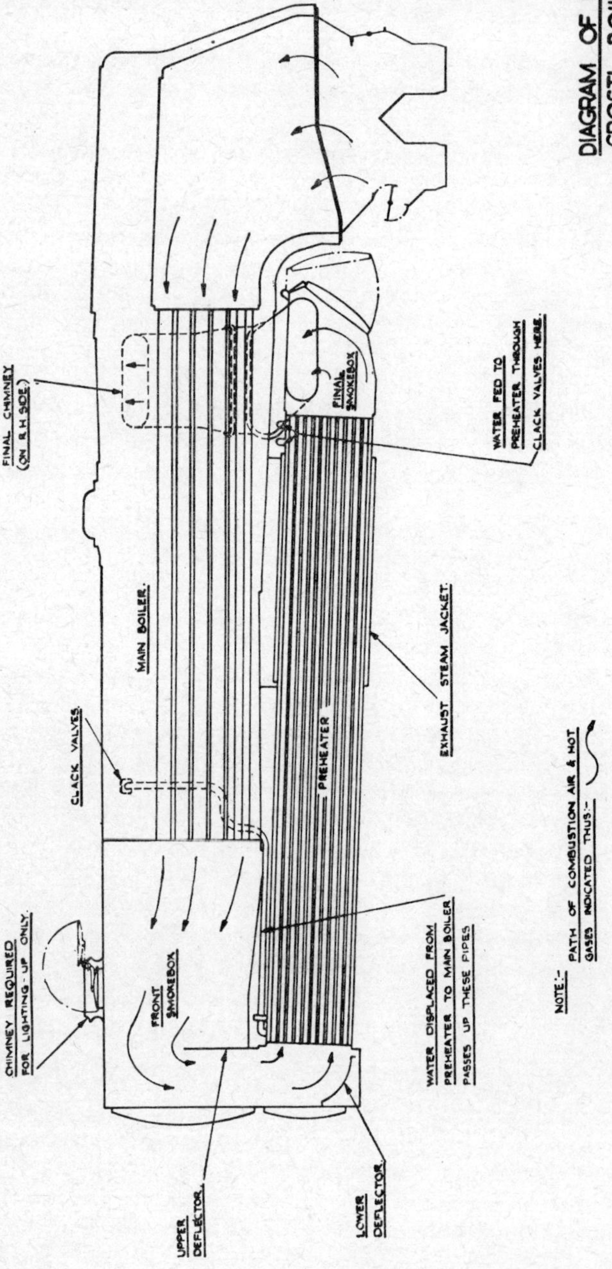

DIAGRAM OF FRANCO CROSTI BOILER.

Fig. 22.

The principle indicated by Fig. 22 was one of contra-flow, whereby the hot gases and cold water flowed in opposite directions through the system, thus permitting maximum extraction of hcat, in that the coldest water was heated by the coldest gases at the remote end of the pre-heater, and the hottest water, which is that in the vicinity of the main smokebox, heated by the hottest gases. There was a debit of course in the increased energy required from the blastpipe and chimney in order to pull the gases through both banks of tubes, but this was mitigated by use of multiple blastpipe tips in line exhausting into a long thin chimney. Although in its country of origin, it was customary to use two pre-heater drums lying along each side of the engine, the British L_2 loading gauge would not permit, and a single drum was provided placed between the frames underneath the main boiler barrel. This drum, sloping downwards towards the rear, had a final secondary smokebox just in front of the firebox throatplate of the main boiler, and the blast chamber and final chimney were located on the right-hand side of the barrel.

Fig. 23 shows these modified locomotives in diagram form with leading dimensions and particulars, and Plate 43 illustrates one of the engines at work as originally completed.

The whole arrangement was the invention of Dott. Ing. Piero Crosti of Milan, a small round man of boundless energy and infectious enthusiasm, who undertook the basic layouts for the application of his system, and close consultation was maintained with him throughout the design, building and subsequent testing. In the course of a visit to Italy on holiday I took the opportunity to see the Crosti-boilered engines in action which were stationed at Venice depot, and I there had the novel experience of leaving my wife in St. Mark's Square to board a "vaporetti" along the canals and to go to work at the depot beyond the central station on the mainland. There 2–6–0s and 2–6–2s of the Italian State Railways were deployed in some numbers, and it was possible to study the various aspects of firing, operating and servicing these boilers. A footplate run to Verona on a 2–6–2 with Caprotti valve gear was an unusual but not very remarkable experience—Italian steam locomotives were not in the forefront of European practice as regards sparkling performance. Later in the same visit I saw the Crosti 2–8–0 freight engines at the Milan depot, and pursued further discussions with the volatile master himself on his home ground. It is of some interest that an intermediary of whom we saw a good deal in the early days of the negotiations was Kalla-Bishop who is well known to all railway enthusiasts for his writings on Italian railway matters.

One result of my visit to the country of origin was to realise to its full horror the effects of omitting the chimney altogether from its usual position on top of the main smokebox. If, as I have previously remarked, the shape of the chimney is one of the most decisive factors in an engine's overall appearance, no chimney at all is the ultimate solecism, and I came back more than determined that such an omission would not occur on our own locomotives. Fortunately, however, the good Doctor actually

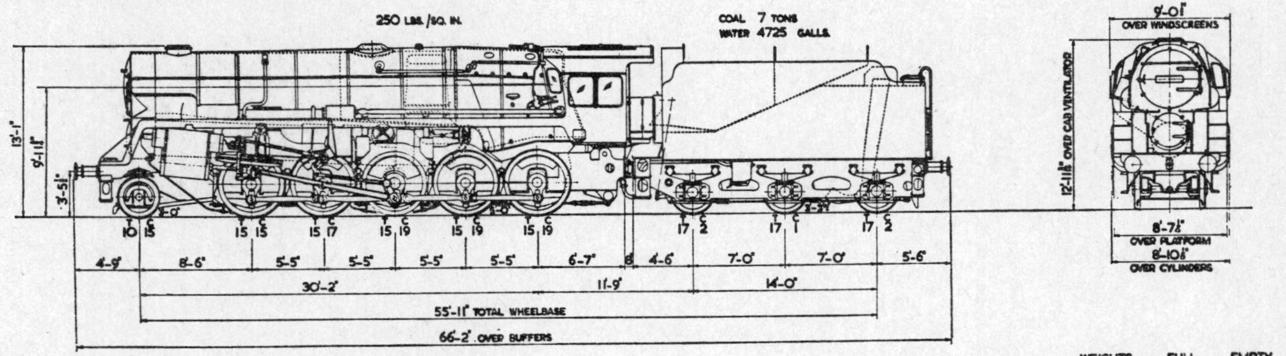

BOILER	
BARREL DIA. (OUTSIDE)	4'-11⅞" INCREASING TO 5'-7⅝"
FIREBOX (OUTSIDE).	7'-5⅝" LONG × 7'-0¼" TO 6'-6¾" WIDE
TUBES	28 LARGE 5⅛" O.D. × 7 S.W.G.
	60 SMALL 2⅛" O.D. × 11 S.W.G.
SUPERHEATER ELEMENTS	1¼" O.D. × 9 S.W.G.
LENGTH BETWEEN TUBEPLATES	16'-5"
PREHEATER	
BARREL DIA. (OUTSIDE)	2'-8⅝"
TUBES	90–2¼" O.D. × 11 S.W.G.
LENGTH BETWEEN TUBEPLATES	19'-3"

HEATING SURFACES	
BOILER	
TUBES	1274 SQ. FT.
FIREBOX	158 SQ. FT.
TOTAL EVAPORATIVE	1432 SQ. FT.
SUPERHEATER	411 SQ. FT.
PREHEATER	
TUBES	1021 SQ. FT.
EXHAUST STEAM JACKET	57 SQ. FT.
FREE FLUE AREA (BOILER)	4·59 SQ. FT.
FREE FLUE AREA (PREHEATER)	2·0 SQ. FT.
GRATE AREA	40·2 SQ. FT.

CYLINDERS (TWO)	20" × 28"
TRACTIVE EFFORT	39,667 LBS.
ADHESION FACTOR	4·48
BRAKE % ENGINE & TENDER	68·4
MINIMUM RADIUS CURVE WITHOUT GAUGE WIDENING	6 CHAINS (OR 4½ CHAINS DEAD SLOW)
BOILER TYPE	B.R.12
TENDER TYPE	B.R.1B.

WEIGHTS	FULL T	FULL C	EMPTY T	EMPTY C
ENGINE	90	4	82	2
TENDER	51	5	23	3
TOTAL	141	9	105	5

SCALE 0 2 4 6 8 10 FEET

B.R. CLASS 9, 2-10-0 CROSTI ENGINE

Fig. 23.

recommended in our case a chimney in the conventional position for the purpose of assisting the draught during lighting-up. For normal running a door which was provided in this chimney could be secured in the closed position by a central dart.

The whole enterprise was the subject of an agreement between British Railways and the Societa Anonima Locomotive a Vapore Franco, in which payment to the latter would be related to the magnitude of the coal savings obtained during tests under specified conditions. The results of these tests are referred to in Chapter VII.

As a footnote to this phase and while initial enthusiasm still rode high a proposal was discussed to follow on by building some Class 5 4–6–0s with this boiler, but nothing came of the matter.

However large a locomotive may be developed, its maximum output with hand firing in this country, whilst retaining a single fireman, is limited by what human endurance will stand, namely around 30,000 lb. of steam for one hour or 24,000 lb. per hour for several hours, these figures diminishing considerably with any but the best grades of coal. Higher outputs the world over, other than those achieved by supermen, are assured by use of mechanical stokers. In order to explore what increased output was attainable on the Class 9 2–10–0s by the latter means, and to ascertain the cost of achieving it, three engines, Nos. 92165, 92166 and 92167 were in 1958 fitted with the American Berkley stoker. Fig. 24 shows these engines in diagram form and Plate 44 the arrangement of the stoker and its controls in the cab. The stoker engine which was mounted on the tender front dragbox was a simple robust machine, infinitely variable in the speed at which it operated the main conveyor screw lying in the trough at the bottom of the coal space. Although graded sizes of coal were required for best results, any larger lumps were drawn past a crushing grid incorporated in the conveyor and the stoker engine was made reversible to deal with possible blockage in the mechanism. After passing up a secondary vertical screw conveyor the coal was delivered on to a distributor plate just inside the firehole whence four separately controllable steam jets directed the fuel to the front and back of the left- and right-hand sides of the firegrate respectively. The control manifold for the above is to be seen on the fireman's side of the cab in Plate 44 and the three small pressure gauges below the larger boiler pressure gauge, indicate on the lower centre gauge the stoker engine pressure, and in the case of the right- and left-hand gauges, a black hand indicates the jet pressure for the back corners, and a red hand indicates the pressure for the front corners of the firebox. Thus an infinitely variable coal distribution was assured in the firebox and unlike some forms of stoker, this one also readily permitted hand firing through the "butterfly" type firedoor in cases of breakdown or abnormal conditions. Here again the arrangement was the subject of most exhaustive tests at Rugby Test Plant and on the line the results of which appear in Chapter VII. The last change from standard so far as the engines themselves were concerned was the fitting of improved draughting arrangements. Minor adjustments made to the

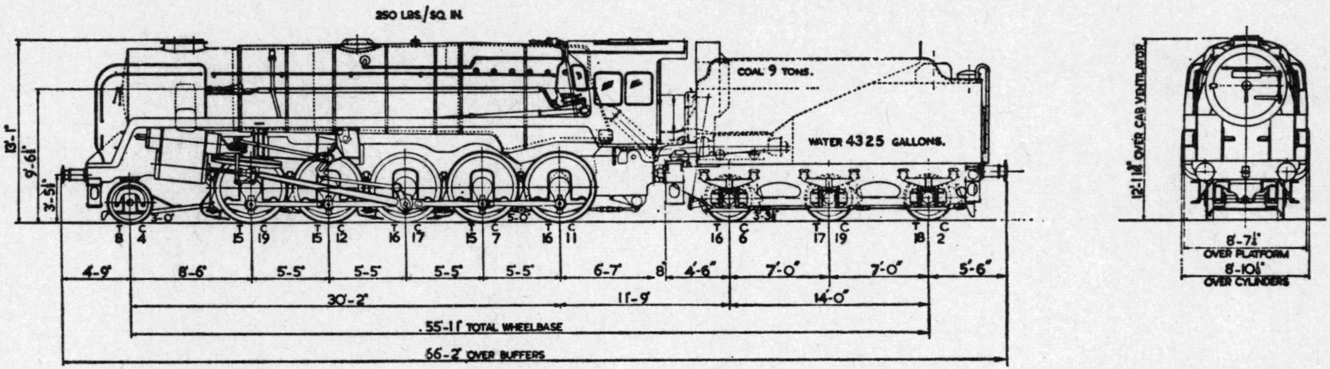

BOILER BARREL DIAMETER (OUTSIDE)	5'-9" INCREASING TO 6'-1"
FIREBOX (OUTSIDE)	7'-5⅝" LONG x 7'-0⅛" TO 6'-6¼" WIDE
TUBES	35 LARGE 5¼" O.D. x 7 S.W.G.
	138 SMALL 2" O.D. x 11 S.W.G.
SUPERHEATER ELEMENTS	1⅜" O.D. x 9 S.W.G.
LENGTH BETWEEN TUBEPLATES	15'-3"
HEATING SURFACES: TUBES	1836 SQ. FT.
FIREBOX	179 SQ. FT.
TOTAL EVAPORATIVE	2015 SQ. FT.
SUPERHEATER	535 SQ. FT.
FREE FLUE AREA	5·49 SQ. FT.
GRATE AREA	40·2 SQ. FT.

CYLINDERS (TWO)	20" x 28"
TRACTIVE EFFORT	39,667 LBS.
ADHESION FACTOR	4·56
BRAKE % ENGINE & TENDER	69·5
MINIMUM RADIUS CURVE WITHOUT GAUGE WIDENING	6 CHAINS (OR 4 CHAINS DEAD SLOW)
BOILER TYPE	B.R. 9
TENDER TYPE	B.R. 1K.

WEIGHTS	FULL T C	EMPTY T C
ENGINE	88 10	79 16
TENDER	52 7	24 13
TOTAL	140 17	104 9

SCALE 0 2 4 6 8 10 FEET

B.R. STANDARD CLASS 9. 2-10-0 ENGINE
FITTED WITH MECHANICAL STOKER.

Fig. 24.

proportions of single chimneys are referred to later, while a number were fitted with double blastpipes and chimneys. One engine No. 92250, was fitted with a Giesl ejector in 1959.

This was one further, and final development of draughting appliances which through the years had progressed from plain circular nozzles through square and clover leaf single variations, with or without different cowl-like attachments as exemplified by Kylala and Kylchap, through double and triple nozzles to multiple jet arrangements. Of the latter the Le Maitre type developed on the French Nord Railway deployed its jets in circular formation in connection with a chimney of abnormally large diameter an arrangement brought over to England by Bulleid of the Southern Railway, and applied to his Merchant Navy and West Country Pacifics, as well as to a number of his predecssors 4–4–0, 4–6–0 and 0–6–0 types. Dr. Adolf Giesl-Gieslingen had spent his earlier days with the Vienna Locomotive Works at Florisdorf, where he had much to do with the impressive 2–8–4 locomotives which were the final expressions in his own country of the philosophy of engine design initiated by the famous Golsdorf. Starting at the time of the design of these engines, Dr. Giesl became specially interested in draughting and he gradually developed the system now associated with his name. In this he devised means of accelerating the exhaust gases before they came into contact with the exhaust steam with the object of reducing the shock loss occasioned by the difference in speed of the steam issuing from the blastpipe orifice, and that of the exhaust gases in the smokebox. To achieve this he finally arrived at a single oblong chimney relatively long in relation to its minimum width, into which multiple nozzles in line, seven in the case of the Class 9, ejected the exhaust steam. The objective of all this, as in the case of its forerunners, was to produce increased draught on the fire for a given exhaust pressure or conversely to produce the same draught at reduced back pressure. The latter of these features in turn could be expected to increase power for a given coal consumption or promote fuel economy at any given level of power output. Fig. 25 shows the arrangement adopted for No. 92250.

As in the case of the Crosti Boiler an agreement was drawn up making payment and subsequent further fitting dependent upon the savings obtained under controlled test on the stationary plant, and how the device fared in this respect will be referred to later.

Apart from the above, a number of locomotives were fitted with straight forward double chimneys and blastpipes—No. 71000 as already mentioned, 80 of the Class 9 2–10–0s, and at a later stage all of the Class 4 4–6–0s allocated to the Southern Region and some of those on the Western Region were altered to take the double arrangement. In spite of the close control of the drawing offices which had been so successful during the bulk design in assuring identical common features, a breakaway occured in the case of these Class 4 engines in that the casting produced at Swindon for their double chimney application was different from that produced at Eastleigh for the Southern contingent. The former of these was as tall

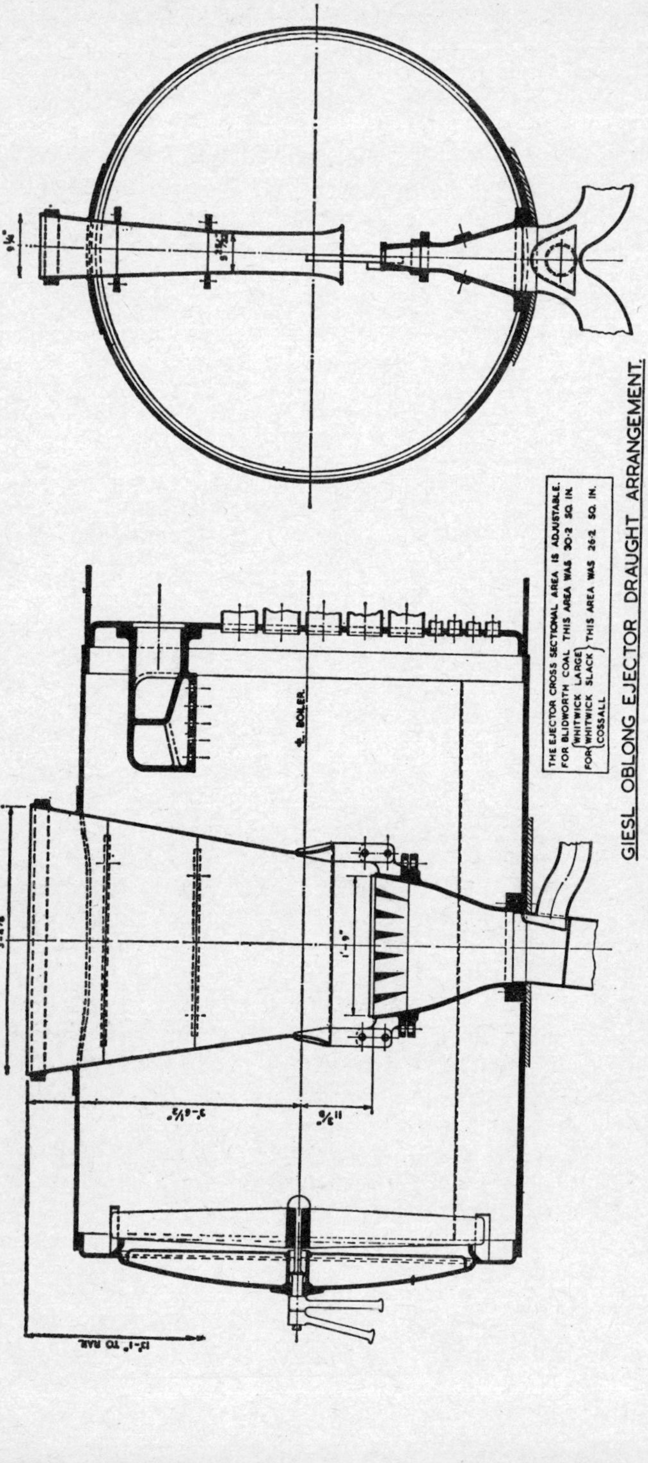

Fig. 25. Giesl oblong ejector as fitted to a class 9 2–10–0.

as the L_1 loading gauge would permit and as seen in Plate 45 produced a very unlikeable and clumsy effect upon the appearance. That fitted to the Southern engines on the other hand was shorter and gave a very harmonious and pleasing aspect to the engines as Plate 46 will indicate. This was a minor and somewhat amusing case where standardisation was caught napping!

Finally in this account of non-standard versions reference must be made to tenders. The original series of all of the tender engines except No. 71000, were fitted with one or other of three basic sizes of tender, all having an inset coal bunker whose object was to permit improved vision to the engine men when running backwards. This was a frequently met feature in Continental steam practice and had been employed for Riddle's 2–8–0 and 2–10–0 W.D. locomotives. Although a very logical arrangement, it was not one which proved to be particularly appreciated by the engine men, possibly because when setting back on to a train, it is the bottom rear corner of the back end of the tender which had to be watched as the leading coach or wagon of the trains was approached, and the inset coal bunkers was no help in this connection. Thus when increased water capacities became necessary, especially for the Southern Region without water troughs, and for service on the Holyhead route and into Scotland where more coal capacity became necessary, the two larger tenders were redesigned with flush sides to the full loading gauge width curved inwards at the top, in the manner formerly applied by the L.M.S. Railway. Not only were water capacities varied to meet operating circumstances, as time went on, but the amount of coal carried was also varied. Provision was made for 9 or 10 tons to be carried in certain cases, but conversely for the shorter runs, the partition marking the end of the coal space was brought forward so as to limit the amount carried to 7 tons. When much more coal is carried than is habitually burned, that at the back of the coal space tends to remain unused for weeks and months as fresh coal is added to the front end, and thus deteriorates in quality so that if and when it eventually finds its way to the shovel bad steaming can result. The varying capacities of all the tenders both inset and flush-sided, other than the initial versions fitted to the engines as built are set out in Table XVI and the appearance of the different tenders can be judged from the various illustrations.

We have now dealt with the inception, design and building of this final series of steam engines for British Railways, and have noted their names and numbers, their appearance and dimensions, and the few departures from standard to which they were subject. It now remains to observe them in service and to examine the nature of their mettle.

VII

Test Results

LOCOMOTIVE TESTING consists in the main of three things, namely:—

(1) Proving the draughting arrangements for efficiency and maximum evaporative capacity.

(2) Establishing the relationship between coal rate, steam rate and power at the cylinders and at the drawbar.

(3) Determing the operating characteristics of the locomotive in respect of range and efficiency.

It was fortunate that by the advent of the B.R. standard locomotives, equipment and procedure was available for the accurate measurement of all these factors. The equipment consisted firstly of stationary test plants at Swindon and Rugby in which the locomotives could be placed on sets of rollers coupled to hydraulic brakes in such a manner that any desired tractive effort could be simulated at the driving wheel rims. Plate 47 shows an engine undergoing such a test, and as it was held physically stationary even although its wheels may be turning at high speed under maximum power, the ease with which conditions could be held constant or varied at will, and the facility with which all measurements could be taken, is obvious. These measurements consisted principally of indicator diagrams at the cylinders, pull at the drawbar recorded on a stationary dynamometer, coal and water consumptions, rate of steam flow to the cylinders, steam and gas temperatures, and such observations as were needed to relate for each rate of steam-production the amount of air supplied, the draught in the smokebox and resistance of the boiler to flow of gases.

The operating characteristics of the locomotive could not be finalised from the results of stationary plant testing alone, since they could not include the aerodynamical and frictional losses always associated with the movement of vehicles over the permanent way. Such testing had, therefore, to be supplemented by a form of controlled road testing which consisted of running the locomotive on the line with normal loads in steps of constant rates of evaporation and combustion to which the power at the drawbar as measured by a dynamometer car (which formed the leading vehicle) was referred. Plate 48 shows this part of a test being carried out, in which indicated power at the cylinders and pull at the drawbar were measured and the locomotive's internal resistance was derived from the difference.

The link between the stationary and the line tests which assured complete uniformity between them, was use of a common steam flow meter, and the firing of the boiler in each case by a method called "Summation of Increments". These devices, which turned first principles of thermodynamics to account in the simplest possible manner and with only the simplest of equipment, were the invention of S. O. Ell in charge of locomotive testing at Swindon, who, with his colleagues at that centre, had since before the war developed a whole philosophy of testing which had taken the art a long step forward from the elementary form of dynamometer car observations which many railways had indulged in since the earlier part of the century.

Since the pressure just below the blastpipe orifice is proportional to the rate of steam flowing through the cylinders, the steam flow meter was simply a means of measuring the former value, a U-tube monometer, duplicated as between locomotive footplate and dynamometer car and calibrated for each individual engine serving to indicate the rate of consumption at any given moment. By selecting the desired rate for a given test, indicated by a moveable pointer on the footplate monometer, the driver was able so to operate his locomotive by manoeuvering regulator opening and cut off as to maintain the level of fluid in the U-tube of the monometer, to correspond with that of the pointer and thus assure a constant steam rate irrespective of road speed or gradients.

The "Summation of Increments" method of firing, which required no special equipment at all, consisted of placing coal at the disposal of the fireman in small quantities or increments of equal weight, sufficient for about five minutes burning. Between each successive increment, the water injected during the increment in order to maintain a constant level of water in the boiler was measured, and the time was noted. The summations of coal fired and water injected were each plotted against elapsed time as the test progressed. The drawing of fair straight lines through each of the plots, verified the uniformity of the test at each given rate of output, and the slopes of the lines established the basic coal and water rates for the test.

These methods were applied to both the Swindon and Rugby stationary tests and to the corresponding line tests which in the former case were usually carried out by the same personnel over a distance of some 76 miles along the main line through Swindon, using trains of empty stock behind the dynamometer car which might attain 20 or more bogie coaches with the largest locomotives under full power. In the case of engines tested at Rugby plant, the road tests were carried out by a separate team from the Derby Technical Office of the L.M. Region, the running usually taking place over suitable stretches either of the Settle-Carlisle line, or of the old G. & S.W. line between Dumfries and Kilmarnock. Since at these latter locations very long trains were an operating inconvenience, relatively short rakes of coaches were supplemented by added resistance from one or more mobile testing units, vehicles like electric motor coaches in reverse, whose traction motors acted as generators, the power developed being dissipated in resistances in a similar manner to that of rheostatic braking on electric locomotives.

Table XVII

Performance and Efficiency—B.R. Locomotives

Locomotive and coal used	*At maximum rate of evaporation*								*At minimum indicated steam consumption*							
	Water Evaporated lbs./hr.	*Maximum I.H.P.*	*Maximum D.B.H.P.*	*Steam Temp. °F.*	*Steam per I.H.P. Hour lbs.*	*Coal per I.H.P. Hour lbs.*	*Boiler Efficiency %*	*Cylinder Efficiency %*	*Water Evaporated lbs./hr.*	*Maximum I.H.P.*	*Maximum D.B.H.P.*	*Steam Temp. °F.*	*Steam per I.H.P. Hour lbs.*	*Coal per I.H.P. Hour lbs.*	*Boiler Efficiency %*	*Cylinder Efficiency %*
B.R. 4–6–2 Cl.8																
S. Kirkby	32750	2630	2100	770	13·0	2·60	50	14·5	22800	1980	1440	710	12·2	1·52	72	15·7
Blidworth	29000	2420	1920	750	12·4	2·55	44	15·0	22800	1980	1440	710	12·2	1·81	68	15·7
B.R. 4–6–2 Cl.7																
S. Kirkby	30200	2220	1800	710	14·4	1·95	68	12·9	19000	1575	1280	655	13·3	1·55	77	14·2
Blidworth	30000	2200	1800	737	14·4	2·30	66	12·7	19000	1575	1280	636	13·3	1·95	72	14·0
B.R. 4–6–0 Cl.5																
S. Kirkby	25000	1760	1420	683	15·0	2·20	64	12·4	18500	1350	1100	635	14·3	1·75	74	13·3
Blidworth	24000	1680	1350	665	14·6	2·20	63	12·8	18500	1350	1100	638	14·3	1·96	72	13·3
L.M. 4–6–0 Cl.5																
Firbeck	22000	1600	1320	681	14·5	2·20	66	12·9	18000	1290	1030	669	14·0	1·85	73	13·6
B.R. 4–6–0 Cl.4																
Blidworth	19600	1420	1120	643	13·9	2·30	61	14·0	17000	1220	950	635	13·7	2·10	68	14·1
Bedwas	19600	1420	1120	643	13·9	2·00	65	14·0	17000	1220	950	635	13·7	1·65	71	14·1
L.M. 2–6–0 Cl. 4 (B.R.)																
Blidworth	17000	1150	970	680	14·7	2·50	63	13·0	12000	830	680	660	14·4	1·85	77	13·3
Bedwas	17000	1150	970	680	14·7	2·20	63	13·0	12000	830	680	650	14·4	1·70	78	13·4
B.R. 2–10–0 Cl.9																
Blidworth	29000	2070	1780	735	14·4	2·20	64	12·6	18850	1530	1225	655	13·6	1·85	73	13·9

Calorific Values BTU's/lb.

S. Kirkby (Yorkshire)	13,800
Blidworth (Midlands)	12,600–12,800
Firbeck (Midland)	13,000
Bedwas (Welsh)	13,800–14,000

Table XVII summarises the results of such tests under the alternative conditions of maximum rate of evaporation and minimum indicated steam consumption.

The end products of all this work were for each locomotive class or variation thereof, a test report or bulletin which set out for each kind of coal used, a series of graphs showing over the whole range of attainable power and speed, indicated and drawbar characteristics, water and coal consumptions, steam and gas temperatures, rates of evaporation, steam—gas—draught and blastpipe pressure relationships, and boiler and cylinder efficiencies. For those requiring more detailed information on the test procedure than this short account can provide, there are two comprehensive papers*† by S. O. Ell and D. R. Carling, in charge respectively of the Swindon and Rugby plants over the period we are now dealing with. All of the B.R. standard tender types were thus tested, except the Class 6, 4–6–2 (72000) and the Class 3, 2–6–0 (77000). In the case of the Class 4, 2–6–0 (76000) and Class 2, 2–6–0 (78000) it was the corresponding L.M.S. types which were originally tested, and since these were in all relevant particulars identical to their B.R. counterparts, the tests were not repeated for the standard version. No tank engines were tested as these do not lend themselves to the required means of handling the coal measurements.

Although Swindon procedure guided the main principles of this large testing programme, useful contributions came from other sources, and Derby and Darlington, each with a long tradition of engine testing, as well as the B.R. Research Department, who were helpful on instrumentation, assisted with different aspects of the work. The whole activity was planned and supervised by an enlarged Locomotive Testing Committee of which Riddles appointed me as chairman. Foundation members besides Ell and Carling already mentioned, were T. M. Herbert, Director of Research for the Railway Executive, R. F. Harvey, Chief Officer (Motive Power) R.E., Dr. H. I. Andrews, Electrical Engineering Department, R.E., C. S. Cocks, L.M. Region, Derby, R. G. Jarvis, Southern Region, Brighton and B. Spencer, E. & N.E. Regions, Doncaster.

There is thus available a large volume of authentic data as to what was designed and built into this series of locomotives. Many of the bulletins have been published by British Railways, and are available to students desirous of following the technical aspects in detail. Here we must confine ourselves to what is of more general assessment and knowledge, and I propose to deal with each class in turn enlarging upon matters of interest in a more general context as we go along.

Taking first the Class 7 Britannias, No. 70005 was selected for initial testing and confirmatory tests were later carried out on No. 70025. The engines were tested as new built, after a suitable period of running in and

* "Locomotive Testing on British Railways", D. R. Carling, *Proc. Inst. Loco. E. Journal* No. 217, 1950.

† "Developments in Locomotive Testing", S. O. Ell, *Pro. Inst. Loco. E. Journal* No. 235, 1953.

the trials themselves took place at intervals between April 1951 and February 1952. Individual tests on the Rugby Plant were of 90 to 140 minutes duration, dropping to 70 minutes at the maximum sustained steaming rate. These periods were the test periods proper during which continuous steaming in stabilised conditions was sustained after a considerable period of "warming up" at the same rate of steaming. The controlled road tests were carried out between Carlisle and Skipton, where, due to the natural limitations of the route, the duration of the tests was shorter, usually 60 to 75 minutes. At the higher rates of steaming, coal consumption exceeded the rate which one fireman could sustain for such periods, and two firemen were employed. The boiler proved itself an excellent steam producer of up to 30,000 lb./hr. with Blidworth coal which was the maximum which the draughting arrangements would sustain on the test plant, although on the road a steam rate to cylinders of 37,560 lb./hr. was maintained for 45 minutes, equivalent to hauling an 850-ton passenger train over the Carlisle-Leeds route at limited load timings. Such timings were only rostered for a load of 405 tons for a Class 7 locomotive in the L.M. Region's working timetables, and so far ahead of this was the test performance that the blastpipe and chimney arrangement was judged satisfactory, and no change to it was found necessary during the lives of these engines. At maximum output blastpipe pressure was $8\frac{1}{2}$ lb./sq. in. and the smokebox vacuum 7.5 in. of water of which passage of air through the ashpan required 1 in., through the firebed 2.5 in., the tubes 2.5 in. and to get the hot gases past the self-cleaning plates and spark-arresting grids, 1.5 in. Boiler efficiency had a very acceptable flat characteristic; 75 per cent at 2,000 lb. of coal per hour, it was 72 per cent at 3,000 lb. of coal per hour, the limit of sustained effort for one fireman at which point 19,500 lb. of feed water was turned into steam. Efficiency was still 66 per cent at 5,066 lb. of coal per hour corresponding to the sustained front end limit of 30,000 lb. of feed water per hour. The actual amount of steam reaching the cylinders was slightly greater than the quantity of feed water evaporated in the boiler due to the action of the exhaust steam injector. Corresponding steam figures were 21,000 lb. and 31,410 lb. respectively. Steam temperature at maximum output was 737°F. and the matching of the tube areas and surfaces as between the small and large tubes in the boiler was shown to be very satisfactory by corresponding exit temperatures of the hot gases from these tubes of 730°F. and 720°F. respectively.

The cylinders, which provided sharp clear indicator diagrams at all steam rates, produced 2,200 i.h.p. at maximum steam output of 30,000 lb. per hour at which point 35 per cent cut-off at $65\frac{1}{2}$ m.p.h. produced a steam consumption of 14.4 lb. per i.h.p. hour. The lowest specific steam consumption was 13.3 lb. per i.h.p. hour at 20 per cent cut-off at 70 m.p.h. when 20,500 lb. of steam per hour (19,000 lb. of water) were being used to develop 1,505 i.h.p. At this point, very typical of what service conditions would call for, cylinder thermal efficiency was 14 per cent which is 68 per cent of the maximum efficiency possible under the Rankine steam cycle on which all the Stephensonian steam engines work. The engine would perform while running with perfect freedom at any cut-off up to 52 per cent, at which value a i.h.p. of 1,920 was attainable at 30 m.p.h., specific steam consumption then being 16.2 lb. per i.h.p. hour. These last figures are worthy of close attention as showing the inherent economy of a free-running modern steam engine in which the whole steam circuit has been opened out according to principles originating from Chapelon. They mean that as between running fast at 70 m.p.h. with 20 per cent cut-off, exerting a tractive effort at the wheels of 8,000 lb. on the one hand, compared with slogging up some gradient at 30 m.p.h. with 50 per cent cut-off and exerting a tractive effort of 23,000 lb. at that speed on the other hand, specific indicated steam consumption rises from 13.28 to 16.2 lb. per i.h.p. hour, an increase of only 22 per cent.

So far as drawbar characteristics were concerned, 1,800 d.b.h.p. on the level at constant speed was attainable at 40 per cent cut-off and 45 m.p.h. At 70 m.p.h. 1,550 h.p. was attained at 32 per cent cut-off. Minimum coal per d.b.h.p. under similar conditions was 2.45 lb. when exerting 1,100 h.p. at the drawbar and using 18,000 lb. of steam per hour. Maximum drawbar thermal efficiency was 8.3 per cent.

To sum up, the cylinders were fully capable of Class 7 duties at an efficiency consistent with other modern locomotives working over the same range of pressure and temperature. The unusually larger boiler in relation to cylinder capacity conferred three potential advantages upon the Operating Department if full use could be made of them. These were:—

(1) The limitation which the draughting arrangements imposed upon maximum boiler capacity, assured high boiler efficiency over the range of working of which one fireman was capable, and even at the highest output, the efficiency did not droop away to 50 per cent and below which could occur where more "efficient" blastpipe and chimney arrangements might push boiler output towards the grate limit where no more combustion air could be absorbed.

(2) The larger grate was able to carry more clinker and ash from poor quality coal before the air supply through the fuel bed became sufficiently restricted to affect steaming adversely.

(3) The larger reservoir capacity both in heat and steam provided a reserve for short periods of heavy demand, such as acceleration up a grade after a check, and to tide over temporarily adverse firing conditions.

On the other hand, with poor utilisation and a lot of standing about during the daily roster, the wide grate would be less economical than a smaller narrow grate in the mere process of keeping the fire alight and the water on the boil in readiness for the next trip.

Turning now to the Class 8, 4–6–2 locomotive No. 71000, which may be thought of as an enlarged "Britannia", it will be remembered that the differences consisted of (a) three cylinders 18 in. × 28 in. instead of two 20 in. × 28 in.; (b) Caprotti poppet valve gear instead of Walschaert piston valve gear; (c) a grate area of 48.6 sq. ft. instead of 42 sq. ft., all other boiler dimensions remaining the same; (d) double in place of single blastpipe and chimney. The starting tractive effort of No. 71000 was nominally 39,080 lb. as against 32,150 lb. for 70000, these values being typical of Class 8 and Class 7 tractive efforts generally. With this locomotive the "B.R. Standards" sought to enter a charmed and exclusive circle consisting of engines to which the heaviest and fastest main-line passenger services were entrusted; Stanier's "Duchess", Gresley's A.4, Peppercorn's A.1, Collet's "King" and Bulleid's "Merchant Navy". In the opening words of the official test bulletin—"Besides having several features not previously incorporated in B.R. locomotives . . . it represents the ultimate stage of development of the steam locomotive in this country."

It can be said right away that the engine portion of the locomotive proved to be superb, and showed a cylinder efficiency unmatched by any other simple expansion engine in the world, of which records are available. On the other hand, boiler performance and efficiency was markedly inferior at the higher outputs, so that the whole locomotive, engine part and boiler together, was unable to match, let alone out-perform, the maximum achievements of its established rivals in the form in which it underwent its trials and subsequent service on the line. This in an interesting case history which merits some examination.

The boiler would not evaporate on the stationary plant more than 29,000 lb. of feed water per hour with Blidworth coal (30,250 lb. per hour of steam to cylinders, allowing for the effect of the exhaust injector). With South Kirby coal, the figures became 32,750 lb. and 34,000 lb. respectively. On the controlled road tests these outputs were hardly attained let alone exceeded, thus showing marginal inferiority to the "Britannia" boiler in this respect. To achieve 30,250 lb. of steam to the cylinders per hour with Blidworth coal, 6,850 lb. of coal per hour had to be fired as against 5,066 lb. in the case of "Britannia". Whereas at the latter rate of firing No. 70000's boiler was still buoyant, showing an efficiency of 67 per cent, and capable of still higher output with an increase in draught, in the case of No. 71000, the grate limit had been reached which no improvement in draught was capable of influencing, and overall boiler efficiency had already dropped away to 44 per cent. This inferiority, hardly perceptible at a firing rate of 3,000 lb. of coal per hour, became very marked as the rate increased, and the resultant lowering of boiler efficiency negatived at the higher rates of working, the excellent performance of the cylinders. This latter feature is exemplified by the ability of the engine to produce 2,440 i.h.p. at 65 m.p.h. at a steam supply of 30,000 lb./hr. as against 2,200 i.h.p. for Britannia. At maximum evaporation of 32,750 lb./hr. and at 90 m.p.h., 2,630 i.h.p. was attainable. Minimum steam consumption was registered of 12.2 lb. per i.h.p. hour at an evaporation of 22,800 lb. per hour and a road speed of 90 m.p.h. when the power output was 1,980 i.h.p. Under these circumstances cylinder efficiency was 15.7 per cent, 86 per cent of that theoretically possible under the Rankine steam cycle. At 70 m.p.h. minimum steam per i.h.p./hr. was 12.2 lb.

compared to 13.3 lb. for Britannia, the latter value representative of best British practice hitherto. The relative cut-offs of the two engines at this point were 12.5 per cent and 20 per cent respectively, and the decrease in specific steam consumption of the former 8.25 per cent. As regards drawbar horsepower, 2,100 on the level at constant speed at 40 per cent cut-off at 40 m.p.h. was the maximum attainable. At 70 m.p.h. 1,850 h.p. was attained at 23 per cent cut-off. Maximum drawbar thermal efficiency was 9 per cent (Britannia 8.3 per cent).

The locomotive as built was thus a near miss. The parts which were good were very good, but the whole thing was let down by this inability to boil water fast. Why was it and what was needed to put it right? The answer is shrouded in mystery, for the bell had tolled for steam, and no more investigation was carried out, since no more such engines were to be built. It is possible to regret that a final measure of investigation was not undertaken which would have completed and rounded off this last phase of the express passenger steam locomotive. In its absence it is only possible to surmise. It will be recalled in the first place that the boiler had become a borderline case for effective steam production, in that the larger grate had reduced the percentage of free area through the tubes to grate area to 14.0 per cent. Although the finger-type rocking grate gave a satisfactory free air space through the bars of 43 per cent, some condition of firebed shape reinforced by a good deal of high frequency vibration at the back end due to the undamped coil springs on the trailing truck, combined together to cause at the higher outputs, considerable disintergration of the fuel on the firebed into small pieces the size of large smokebox ash, a condition characteristic of conditions near the grate limit. In such circumstances, when large accumulations of these pieces occurred, not only was a stream of unburned particles swept through the tubes and up the chimney by the action of the draught, but the air supply through the grate was seriously impeded by the "packing solid" of the incandescent fuel. This explanation leaves many question marks however. Each separate item which seemed to contribute to this debacle, appeared so normal. The double chimney draughting appliance was effective and efficient, producing a smokebox draught of 8 in. of water for a blastpipe pressure of 5.6 lb./sq. in. The percentage of free area through the tubes to grate area although less than for Britannia, was better than on the Class 9, 2–10–0, which proved an excellent steamer. Air passage through ashpan and grate did not seem deficient by an known criterion, while the fact that coil springs were used on the trailing truck does not seem to have caused such marked effects upon "Merchant Navy" steam production because of undue vibrations. We shall never know the true answer now. All that can be said is that with such modifications as would have permitted an economical evaporation of up to 40,000 lb./hr. with best coal, this engine could have been a lively challenger to even the best of its regional brethren.

Even so, the engine as built showed no overall inferiority at coal rates up to 3,000 lb. per hour, and it is interesting to digress for a moment and

Table XVIII

Performance and Efficiency of Various Locomotives at Constant Firing Rate of 3,000 lbs. of Blidworth Coal Per Hour

Locomotive		Water Evaporated lbs./hr.	*Indicated Values*						*Efficiency*			*Drawbar Values*				
			Maxim I.H.P.	At M.P.H.	At Cut-off %	Steam Temp. °F	Steam per I.H.P. Hours lbs.	Coal per I.H.P. Hour lbs.	Boiler	Cylinder	Randane	Maximum D.B.H.P.	At M.P.H.	At Cut-off %	Steam per D.B.H.P. Hour lbs.	Coal per D.B.H.P. Hour lbs.
B.R.	4–6–2 Cl.8	20000	1750	90	6	680	12·2	1·72	70	15·7	86	1260	35	22	▼	▼
	4–6–2 Cl.7	19500	1620	65	22½	698	13·3	1·95	72	14·0	68	1240	35	35	16·0	2·44
	4–6–0 Cl.5	19500	1480	75	27	643	14·4	2·00	70	13·2	68	1180	37	37	16·7	2·53
	4–6–0 Cl.4	18700	1350	65	27	640	13·8	2·20	63	14·0	76	1060	35	39	18·0	2·80
	2–10–0 Cl.9	21000	1600	65	22	695	13·7	1·89	73	13·4	67	1360	30	32	15·5	2·30
L.M.	4–6–2 "Duchess"	22000	1480	50	17	620	14·8	2·02	76	12·9	61	1290	35	—	—	—
	4–6–0 Cl.5	20000	1480	70	28	675	14·0	1·95	69	13·4	68	1200	37	39	16·5	2·40
E.N.E.	2–6–2 V2	21000	1615	80	17	680	13·4	1·85	74	14·3	73	1220	40	30	17·0	2·50
S.R.	4–6–2 "Merchant Navy"															
	Original	21500	1300	55	*	610	16·0	2·35	70	10·0	56	1100	40	*	21·0	3·00
	Rebuilt	24000†	1700‡	90	17	640	14·2‡	1·81‡	74	—	—	1250	40	24	—	—
W.R.	4–6–0 "King"															
	Blidworth Coal	21500	1630	80	18	660	13·9	1·84	74	13·9	68	1210	33	28	18·7	2·48
	Markham Coal	23600§	1770	80	20	670	14·0	1·70	71	13·8	68	1340	35	30	18·6	2·24

Notes: *No Stable cut-offs were established.
†Only Welsh Bedwas coal used on these tests (13800 BTU's per lb.)
‡Engine not indicated. I.H.P. values estimated from D.B.H.P. figures
§Markham Best Welsh coal. Calorific value 14510 BTU's per lb.
▼Values not recorded in test bulletin.

examine what a variety of different locomotives were capable of achieving at this rate of firing which was representative of what a good fireman could sustain continuously over several hours. Table XVIII will repay study for it brings to light some interesting facts.

The test figures for all of these engines were taken when using Blidworth coal or its close equivalent, except in the case of the rebuilt "Merchant Navy" which was only tested on Welsh coal of second grade, and the second entry for the W.R. "King" where first grade Welsh coal was used. While the figures for maximum i.h.p. are comparative one with another, as are the figures for maximum drawbar h.p., the relationship of the former to the latter is not consistent amongst the various engine classes due to the differences in shape of the various horsepower curves over the whole range of working. Finally the figures for coal per drawbar h.p. hour shown in the last column, refers to conditions under constant speed on the level, and are not to be compared to the often quoted results of conventional dynamometer car tests at variable rates of combustion and evaporation. This said, attention is drawn to the following:—

(1) As the size of locomotives in the same family of design decreases, as in the case of the first four B.R. standard locomotives in the table, a decreasing amount of power is obtainable from the given quantity of fuel. This is because the smaller engines are working progressively further from their most economical point and nearer to the limit of their capacity. In particular, boiler efficiency is dropping from engine to engine. Thus there is advantage in using the largest available engine for the heavier duties, rather than to work a small engine harder. This is an aspect of locomotive utilisation which has to be balanced against the low drawbar efficiency of a large engine when it is called upon, as part of its roster, to haul a very light train.

(2) The very close resemblance between all the values for the B.R. Class 7 and the E.N.E. V2 Class 2–6–2, an engine which epitomises all the best Gresley features. Although his A4 Pacific was never put through its paces on a test plant, notwithstanding a token appearance of one of these engines for the opening of the Rugby Testing Station, it is probable from an inspection of its design and proportions that it would have shown very comparable figures to the V2 class.

(3) The close agreement between the figures for B.R. Class 7, 4–6–2 and B.R. Class 9, 2–10–0. These values give some indication of why the latter turned out to be such an excellent express passenger engine on the few occasions it was given the chance.

(4) The even closer identity between the B.R. and L.M.R. Class 5, 4–6–0s. This is referred to again later in this chapter.

(5) The serious inferiority of the S.R. "Merchant Navy" in the form in which it was first built. The indeterminate and random valve events which were a feature of this engine resulted in so low a cylinder efficiency, that to equal the power output of comparable

Class 8 engines, which it could certainly do, a great deal more fuel had to be burned, which in turn led to a further debit due to lower boiler efficiency.

(6) The notable improvement which resulted from restoring the valve gear of the "Merchant Navies" to conventional Walschaerts.

(7) Surprisingly the L.M.R. "Duchess" does not show up outstandingly in this table. The relatively low superheat has much to do with this, the lower steam temperature being due to strangulation of the hot gases in threading their way along element filled super-heater tubes of 19 ft. 3 in. in length but only $5\frac{1}{8}$ in. in diameter. On the other hand, the boiler of this engine was a magnificent steam producer, capable of evaporating 40,000 lb. per hour if the highest grade coal were used, a figure higher than that attained by any of the other Class 8 engines over a duration of a hour or more.

(8) Perhaps most interesting of all is the position taken up by the W.R. "King" class after modification to embody high degree superheat whilst retaining single chimney and blastpipe. It now lines up with remarkable closeness to the performance and efficiency potential of its counterparts of entirely different ancestry and detail design, thus once again emphasising the common level of attainment which had been reached by the different streams of British locomotive design.

Put another way, steam at 220-250 lb./sq. in. superheated to 650-700°F. supplied through modern long lap valve gear to cylinders and valves of modern design could hardly help but produce roughly the same results, so long as combustion conditions remained buoyant. The figures for the "King" also show how little it was handicapped, other than in the quantity of coal burned, when having to use hard coal, bearing in mind that Blidworth was a second-grade product. Had the engine been tested with Grade 1 South Kirby coal, the differences would probably have been only marginal compared with best Welsh.

The tests carried out with the Class 5, 4–6–0 engines contained two features of especial interest. The first was obviously to ascertain how, if at all, the changes which had been made to cylinders, valve gear and wheel diameter, would vary performance and efficiency in comparison with the similar engines of Stanier's design on the L.M.R., on which the new engines were so largely based. The second was the introduction of some variations in the draughting arrangements from which we all learned something new.

No. 73008 was selected for the tests at Rugby and subsequently between Carlisle and Skipton, where for the higher rates of working two of the mobile test plant vehicles were again used to supplement the train of ordinary passenger coaches. The engine was originally designed with a $5\frac{1}{8}$-in. diameter blastpipe, following in all respects the precepts of S. O. Ell, but it was soon ascertained on the stationary plant that fully effective through the lower and medium range of working, it would not permit a greater maximum evaporation than 26,310 lb. of steam (25,000 lb. of water) with South Kirby coal, which was very reasonable for the size of the boiler, but with Blidworth coal only 19,150 lb. of steam (18,000 lb. of water) which was insufficient. This loss in evaporation (27 per cent) was very much more than could be accounted for by the lower calorific value of the Blidworth coal (8 per cent) and showed that the smokebox vacuum generated by the blastpipe pressure were unable to punch its way through the greater resistance to the flow of air arising from the various other features of the lower grade coal. A minor modification to the grate by substituting plain grid type firebar elements in place of the more complex Hulson type, improved air space through the bars from 32 to 37 per cent of grate area with an improvement to maximum evaporation of 1,000 lb. of water per hour. Careful study of the front end arrangements indicated that a reduced blastpipe orifice might be beneficial, leaving the chimney proportions as they were, and a further series of tests was mounted at Rugby with No. 72030 having these features. Evaporation of water with Blidworth coal was increased from 19,000 lb./hr. with the original $5\frac{1}{8}$-in. cap and the improved grate to 22,000 lb./hr. with a 5-in. cap, and to 24,000 lb./hr. with a $4\frac{7}{8}$-in. cap, and the latter was quickly applied to all engines in the class. With the largest cap, maximum output had been reached when a blastpipe pressure of 4 lb./sq. in. had produced 5 in. of smokebox vacuum. With the smallest cap, 9 lb./sq. in. produced 9 in. of vacuum. What we learned, rather to our surprise at first, was that this change in draughting conditions with a single chimney was obtained at no loss in boiler efficiency or increase in steam consumption per 1 h.p. hour throughout the range of working. The reasons for this were, first, that the higher draught had raised the temperature of combustion and of the gases entering the large tubes to such an extent that the admission steam temperature was raised from 660°F. to 708°F. at maximum output and proportionally lower down the power scale. This had counteracted almost exactly the loss in efficiency due to the increase in back pressure. Secondly the reduction in blastpipe cap area still left the velocity of steam through the orifice below the acoustic velocity or the sonic speed in steam. Swindon experimental work had shown that so long as this condition remained, the effectiveness of the blastpipe and chimney arrangement as an ejector was unimpaired. These particular test results throw a fascinating afterlight on the age-old practice of placing "Jimmies" in the blastpipes of steam-shy engines. As is sometimes the case, the instinct of those old shed stalwarts who solemnly fixed a piece of thick wire across the blastpipe cap of every engine of certain classes returning from the workshops, was not so far out after all.

As regards the general test performance of these engines and how it compared with their L.M.R. sisters, Tables XVII and XVIII cover the test results under various conditions of working. Close as these figures are between the two kinds of locomotive, there are some external circumstances bearing on the small differences which have nothing to do with the inherent design. For example the entry in Table XVII for the B.R. engine with Blidworth coal is based upon the $4\frac{7}{8}$-in. diameter blastpipe which eventually became standard. The figures for South Kirby coal are from tests made with the original $5\frac{1}{8}$-in. blastpipe. No measurements were made with this latter coal after alteration to the blastpipe, when it is to be assumed marginally better figures would have been obtained.

In the case of the L.M. engine, the Blidworth Colliery, at the time the tests were made was producing coal somewhat variable in quality, and it was decided for consistency of results amongst themselves to substitute another coal "Firbeck" of the same 2B grading but of rather better clinkering properties, and of a slightly higher average calorific value, 13,000 B.T.U.s per lb. instead of a usual 12,600 to 12,800 B.T.U.s per lb. for Blidworth. Bearing in mind these factors, the following seems to be a fair summing up of the relative abilities of the two engines:—

(1) There seems no discernible advantage or disadvantage in the 19-in. diameter cylinders, $1\frac{11}{16}$-in. steam lap, $7\frac{11}{16}$-in. valve travel and 6 ft. 2 in. diameter wheels of the B.R. engine, compared to the $18\frac{1}{2}$-in. diameter cylinders, $1\frac{1}{2}$-in. lap, $6\frac{1}{2}$-in. travel and 6 ft. 0 in. wheels of the L.M. type.

(2) The L.M. engine had the same "ELL" chimney and blastpipe proportions, but the latter remained at $5\frac{1}{8}$-in. diameter similar to the original B.R. engines. Excellent engine as it was, the L.M. engine could possibly have become even better by a similar sharpening of the blast.

(3) The figures set down in the table indicate a slight advantage to the L.M. engine in cylinder efficiency and in steam consumption. It is impossible to judge whether this advantage would remain had all of the tests been made with engines in identical condition of draughting arrangements and quality of coal. It is sufficient to say that for all practical purposes the potential of the two engines was identical.

One of the biggest regrets I have as regards all of this testing work, was that no tests were made with one of the 30 B.R. Class 5s fitted with Caprotti valve gear. Bearing in mind the exceptionally good cylinder performance and efficiency of No. 71000, and having regard to the very good boiler on the 4–6–0 engine, it is probable that quite outstanding results might have been obtained. Here again, however, night was falling upon all steam activities, and spending money on such further tests was not viewed favourably as under modernisation there was to be no further requirement for locomotives of this type. For the same kind of reasons these engines seldom found their way on to first class main-line duties where their potential supriority over the "standard" engine might have displayed itself in daily working.

The Class 4 engines 4–6–0 and 2–6–0 were both tested at Swindon, the former being No. 75006 and the latter No. 43094, an L.M.R. engine which in every feature affecting performance and efficiency was identical with the 76000 series of B.R. engine as built. The tables show how these smaller engines stood up well to the levels of efficiency of their larger brethren. There is no easy explanation as to why the Class 4, 4–6–0 gave better cylinder efficiency and steam consumption figures than did the Class 5 engines, nor why these figures were not equally good on the 2–6–0 engine having regard to the considerably higher steam temperature due to the latter engine having tube ratios more favourable to high superheat. An interesting feature of these latter tests is, however, to observe the effects of using Midland and Welsh coals, both of Grade 2, and of 12,560 and 14,050 B.T.U.s per lb. respectively. Not a single pound more of water was evaporated, not a horsepower more was produced at cylinders or drawbar, and no improvement in steam consumption or cylinder efficiency was obtained at any output by the use of the Welsh coal. The sole advantage was a reduction in the coal used per hour and per horsepower, closely proportional to the difference in calorific values. At maximum output, the 4–6–0 burned 13 per cent less coal per hour and the 2–6–0 12 per cent less. The price of Bedwas coal was, however, just 14.8 per cent more than Blidworth. Against this background, together with what has already been said about the "Kings", it is difficult to retain much sympathy for the Western Region people who used to complain that their engines were dreadfully penalised in comparison with others when they were made to run on anything but Welsh coal.

At 19,600 lb. of water evaporated per hour, the B.R. Class 4 4–6–0 was still a long way from its grate limit, and its boiler efficiency was thus safeguarded while the single chimney front end exerted its restraining influence. However, to provide a greater margin of performance on certain duties where axle weights prohibited Class 5 engines, a double blastpipe and chimney was developed at Swindon for the Class 4s as already mentioned in the last chapter. This proved able on check testing to evaporate 22,400 lb. of water per hour, and although not actually verified would probably give around 19,800 lb. at a firing rate of 3,000 lb. of Blidworth coal per hour. At any higher rate of working than the latter, the boiler efficiency would of course drop to uneconomical levels as the grate limit was approached.

The little Class 2, 2–6–0 was tested in its L.M.R. form by Swindon in 1950 long before the B.R. version with its quite minor variations took the road. The purpose of these trials was mainly to tune up Ivatt's original design by improvements to the draughting arrangements. No indicating was undertaken, so that results comparable to those shown in the tables were not obtained. The engine was, however, run on the plant and under controlled road tests. A change from the L.M.S. chimney proportions of 1 ft. 1½ in. diameter at the choke and 1 ft. 5 in. at the top, to 1 ft. 0 in. at the choke and 1 ft. 2 in. at the top, together with slight adjustments in vertical height but retaining the 4⅛-in. diameter blastpipe cap in each case, raised the maximum evaporation from 9,850 lb. of water per hour to 14,000 lb. using Blidworth coal. As finally adjusted, this gallant little engine showed itself on the road trials able to lift 15 coaches weighing 455 tons up nearly five miles of 1 in 300 between Little Somerford and Badminton, at a minimum speed of 40 m.p.h. At a steam rate of 13,378 lb. per hour it was using coal at 2,501 lb. per hour. In these conditions it was exerting 625 drawbar h.p. at 42 per cent cut-off, the coal per d.b.h.p. hour being 3.6 lb., steam temperature was 585°F. and boiler efficiency 55 per cent.

Finally we come to the B.R. 2–10–0 in its various forms. As originally designed with single blastpipe 5⅜-in. diameter, the draughting was found inadequate and a reduction of the orifice to 5¼ in. sufficed for a maximum evaporation of 29,000 lb. of water per hour which with the assistance of the exhaust injector supplied 30,544 lb. of steam to the cylinders at a consumption of Blidworth coal of 4,655 lb. per hour. While boiler efficiency held up very well up to this rate of working, the shallower grate and reduced firebox volume in comparison with Britannia, as well as the less favourable ratio of free area through tubes to grate area, combined to make good combustion at high output more tricky, and as subsequent tests for other purposes proved, the engine was in these circumstances working near to the limit of adequate admission of air through ashpan and grate. The engine part, as might be expected, gave results almost identical with those of Britannia which indeed was not to be wondered at, having regard to the close correspondence of the thermal range over which the two engines worked and the similarity in cylinders and valve gear. There is no need to repeat, therefore, the principal figures which are set out in the tables. The main series of tests were carried out at Rugby, and by the Derby dynamometer car and Mobile Test Unit operating this time as far afield as Hurlford on the old G. & S.W. main line. With due regard to wheel size, none of these trials were run at higher speeds than 65 m.p.h. and there are unfortunately no test results to illuminate working conditions when the engines were running up to 90 m.p.h. as they proved themselves perfectly well able to do in ordinary traffic during their subsequent lives. Nevertheless as the adjoining tables attest, there was ample technical support for the excellent versatility of which they quickly proved themselves capable in service.

The variants to the Class 9, 2–10–0s referred to in Chapter VI were subjected in their turn to tests, specially designed, in conjunction with the protagonists of the special features, to bring out the particular value, if any, of the respective modifications. Thus the Crosti tests were the subject of an agreement drawn up at the end of March 1953 between Farina of the Franco Company, Crosti the inventor and the Railway Executive, the substance of which was that comparative tests at Rugby Plant and on the line, should be made with a Crosti and with a Standard 2–10–0, each to be certified by the firm that they were in all respects satisfactory for the purpose from their point of view. The tests themselves, which were to be under the joint supervision of both parties, were required to determine the coal consumption per drawbar horsepower at speeds of 20, 30, 40 and 50 m.p.h. each at a constant rate of evaporation of 16,000 lb. of feed-water per hour using Blidworth coal. The average of these four values was to be taken as the coal consumption of each of the test locomotives on which savings were to be determined. The royalty to be paid to the firm per locomotive, was to be on a sliding scale, 100 per cent of the agreed figure if the savings reached 18 per cent, nothing if the savings were only 12 per cent or less, and pro rata in between. The firm had every confidence that the savings would earn it a substantial royalty based on their experience with Italian engines.

The ten Crosti engines were turned out new in 1955, and the tests took place between June and November in that year. The Franco Company appointed Dott. Ing. de Paolis of the Material & Traction Department of the Italian State Railways, as their representative, who was to agree

Table XIX

Actual Test Results—Crosti 2-10-0

		Crosti		*Standard*	
Engine Number		92023		92050	
From the Stationary Test Plant:—					
Steam lbs/hr.		16,000		16,000	
Type of Injector		*Live*	*Exhaust*	*Live*	*Exhaust*
Water lbs/hr.		16,000	14,880	16,000	15,320
Coal lbs/hr.		2,018	1,852	2,162	2,050
From Tests on the Line:—					
Drawbar H.P. at M.P.H.	20	925		976	
	30	933		983	
	40	882		941	
	50	764		839	
	Average	876		935	
Combined Results:—					
Type of Injector		*Live*	*Exhaust*	*Live*	*Exhaust*
Coal lbs/DBHP/hr. at M.P.H.	20	2·179	2·002	2·215	2·100
	30	2·163	1·985	2·199	2·085
	40	2·288	2·210	2·298	2·179
	50	2·641	2·424	2·577	2·443
	Average	2·318	2·128	2·322	2·202
Percentage Savings					
Both engines with live steam injectors				0·5%	
Both engines with exhaust injectors				3·35%	
Crosti with exhaust injector / Standard with live steam injector				8·35%	

Note:—These savings are the actual obtained on test. The slightly different figures quoted in 'Locomotive Panorama' Vol. 2, were taken from the official test report signed by Paolis and Carling in which some allowance was made for the differing weights and internal resistance between Nos. 92023 and 92050.

procedure, attend the trials and sign the final test report, and Crosti himself was also an active participant in all that went on. The high hopes of all parties unfortunately failed to mature for a complex of reasons enshrined in the voluminous correspondence and records, of which it is only possible to give a short prècis here. Table XIX sets out the actual results obtained, while Table XX portrays what might have been expected had things been other than they were. At the beginning some adjustments in the blastpipe and chimney proportions proved necessary, but those finally chosen, although not necessarily the optimum, permitted maximum evaporation rates equal to those of the standard engine. As was to be expected, the boiler efficiency of the combined Crosti boiler and preheater was higher over increasing firing rates, and did not fall below 72 per cent efficiency at 4,500 lb. of coal per hour. Steaming and general performance were good and it was conceded that the tests had been carried out with scrupulous fairness. The agreement had been rather imprecise as to what was to be compared with what as regards use or otherwise of the exhaust steam injector, but on any basis, even the most favourable one of use of this instrument on the Crosti compared with live steam injectors only on the standard, the results fell far below the minimum which would earn any royalty for the firm. Naturally an intensive probe was set on foot as to what had gone wrong with the bright expectations.

In the first place the firm were unlucky in that No. 92050 was chosen in all good faith as the "control". This happened to be one of those engines in which everything was at the top of its form, whereas had No. 92013

Table XX

Hypothetical Results—Crosti 2-10-0

	With Live Steam Injectors	*With Exhaust Injectors*	*Crosti only with Exhaust Injectors*
1. Official Test Results Engines 92023 and 92050	0·5%	3·35%	8·35%
2. Had 92013 been control Engine instead of 92050	4·5%	6·15%	12·45%
3. Had agreed steam rate been 22,000 lbs/hour instead of 16,000	6·6%	8·1%	14·65%
4. With equal co-efficients of heat transfer for both systems (Chapelon's estimates) Engines 92023 and 92050 at 16,000 lbs/hr.	3·5%	6·65%	11·45%
5. As 4. above—Engines 92023 and 92013	8·1%	9·15%	15·65%
6. As 5. above but at 22,000 lbs/hr.	10·45%	11·85%	18·1%

been selected, a more ordinary engine which had some time previously also been tested at Rugby, results more favourable to Crosti would have been obtained, although still not sufficient to earn the company anything appreciable. Then the course of the trials had indicated that the benefits of the system became more marked as the engine was worked harder. An evaporation of 16,000 lb. of water per hour had been chosen as representative of average heavy freight working in this country, corresponding to a firing rate of 2,000 lb. of coal per hour. At the maximum continuous rate for a single fireman of just over 3,000 lb. per hour, an evaporation of 22,000 lb. was obtainable, and had this rate of working been selected for the comparison, the results would have been still better, although not by any very great amount.

So concerned were the firm with all this, that they called in no less a celebrity than Andre Chapelon himself to go over every detail of the test recordings with a fine tooth comb, in order to seek out the nigger in the woodpile. A voluminous report was forthcoming replete with the most erudite calculations and formulae. Chapelon found that while the difference in coefficient of heat transfer between the ordinary boiler and the boiler proper of the Crosti was not appreciable, the value obtained for the preheater was much lower than that of either boiler of the two test locomotives, or for other Crosti applications in Italy and Germany. Had the preheater given its expected results it was calculated that overall boiler efficiency would have been materially increased. Chapelon concluded that the preheater of No. 92023 must have been seriously scaled up inside to have caused this diminution of the expected heat transfer but this is a little difficult to accept as this engine, supplied with softened water and subject to the laid down regime of blowdown and washing out, had only run 80 miles from new when first received at Rugby, and had accumulated but 10,000 miles by the time the trials were over. The point was never resolved because there seemed no purpose in gutting the preheater for internal examination when the tests ended, and by the time Chapelon reported over two years later, the then condition was of course by no means representative of what it had been in November 1955.

It seems clear that the firm had been rendered over optimistic by the scale of improvement achieved in Italy where the type of locomotives fitted, 2–6–0s, 2–6–2s and 2–8–0s were of far from modern design, and where such locomotives had neither feed water heater or exhaust steam injector in their normal condition. It would further seem that about 12 per cent was the maximum saving of which the system was capable, like for like, in modern idiom, but of this, under the conditions of the agreement only 3½ per cent was realised. Two more items require brief reference before the Crosti story is over. Outside the agreement and as an act of grace, supplementary tests were carried out later on the line in order to examine the claim that the heat stored in the preheater drum could enhance efficiency when starting away after lengthy stops. Such was proved to be the case, and a saving of 3½ per cent became enlarged to 8 per cent under these circumstances. The second point is that any savings, great or small, were counteracted by increased maintenance costs arising from the severe corrosion which attacked smokebox and chimney in service. This arose from deposit of sulphuric acid on the plates when the products of combustion became reduced to a critically low temperature during their passage through the preheater tubes, a condition specially likely to obtain during lighting up. No cure was discovered for this at the time, but persistence might have found a solution had the engines continued as built. The engines were highly unpopular due to exhaust from the chimney beating down into the cab in a cross wind (Plate 43) a condition which subsequent fitting of deflector plates alongside the chimney did little to mitigate, and the whole ten of the series eventually had their drums removed and operated thereafter as normal engines. A small ex gratia payment to the Franco Company, without prejudice, eventually concluded this tangled story.

Except for removal of the manual labour, the fitting of mechanical stokers to the three Class 9s did very little for them, thus underlining the world-wide experience that stokers were hardly worth while for grate areas below 50 sq. ft. No. 92166 was the subject of partial tests at Rugby in 1959, exploring only those aspects which were directly affected by the stoker firing; all other aspects had been covered by the tests of these engines in original form as set out in B.R. Bulletin No. 13 published in that year. No greater evaporation rate proved possible than with hand firing, for at 29,000 lb. of water per hour, 6,000 lb. of coal were being consumed by the stoker, in place of 4,750 lb. with hand firing. This larger quantity was on the very limit of what the grate could digest, and indeed it was only obtained after the blastpipe diameter had been reduced from 4 in. to $3\frac{7}{8}$ in. (double nozzles), after the shape of the grate had been altered to achieve a uniform slope for the firebars, and after additional holes had been cut in the front and back of the ashpan in order to admit more air to the underside of the grate. The increase in the quantity of coal consumed above that appropriate to the quality of the fuel and its rate of firing, was due to the production of fines resulting from the passage of the Blidworth small cobbles along the screw conveyors. The steam jets directing the coal on to the firebed had thus to cope with 70 per cent of the delivery to them in the form of cobbles up to 6 in. size and 30 per cent in the form of dust and pieces less than 1 in. Jet pressures suitable for delivery of the intended size of cobbles were too high for the fines which were either carried away up the chimney as unburned fuel, or became piled up in a solid bank under the brick arch thus artificially reducing the active grate area. As a result, boiler efficiency was reduced by from 9 to 12 per cent over the working range when using the stoker.

Since one of the claims for a stoker is to make possible use of lower grades of coal, tests were also carried out with coal taken direct from the coaling plant at Rugby Motive Power Depot, representative of that supplied normally to freight locomotives in the area at that time. It was a mixture of briquettes, ovoids (a form of small briquette) and a little Grade 3 coal with a calorific value of 12,250 B.T.U. per lb. Also tried was Whitwick Hard Slack consisting mainly of dusty slack with some pieces up to 1 in. and containing some pyrites and shale, calorific value being 9,810 B.T.U. per lb. Use of the former mixture increased the coal consumption by 10 per cent above that of stoker-fired Blidworth cobbles, while the Whitwick slack proved unburnable in any stable and economical manner, and it was clearly unsuitable as a locomotive fuel under any circumstances. Steam to the stoker engine varied over the range of output from 100 to 300 lb. per hour and steam to the distributing jets, from 730 to 800 lb. per hour, giving an overall steam consumption for the whole process of from 830 to 1,100 lb. per hour.

The best that could be said of the stoker in this application was that it was capable of sustaining a steady output of 21,000 lb. of steam per hour for several hours on end, at a rate of coal burning of 3,500 lb. per hour which was rather beyond what even the best fireman was capable of. The increase in coal used was, however, some 16 per cent above that of hand firing, and although suitable duties were sought for the three stoker-fitted engines on the Birmingham to Carlisle run, no actual freight duty was found so free of slacks, speed restrictions and delays as to prove beyond the capacity of hand firing, or to exploit the ability of the stoker to maintain high steady outputs. If the only objective was to get maximum tonnage trains over the road regardless of coal cost, the stoker might have justified itself here, but American type of operation was impossible in this small island with its complex railway system and rising coal costs.

A further series of tests were carried out at Rugby, also in 1959 with engine 92250, whose fitting with the Giesl oblong ejector was described in Chapter VI. As in the case of the Crosti, the objective of the tests was the subject of a carefully worded prior agreement with the firm of Schoeller-Bleckmann of Vienna who manufactured the ejectors under Dr. Giesl's patents. Two principal claims were made, first that there would be a considerable reduction in back pressure with a consequent increase of cylinder power for a given steam rate; secondly that low-grade coals, not hitherto in use or deemed useable as locomotive coals, could be burned economically as a result of this equipment. The Coal Board was at this time particularly anxious to find a market for its low-grade coals, and it was on this second objective that the agreement mainly concentrated. The comparison was to be made on the basis of a standard engine working at 40 m.p.h. at a steam rate of 18,000 lb./hr. with Blidworth coal, to serve as the control. The low-grade coals burned with the Giesl ejector at the same rate were to demonstrate money savings of not less than 7½ per cent over the cost of the Blidworth coal. There was a further condition that the boiler should be able to produce not less than 21,000 lb. of steam per hour with any of the coals used. Three grades of coal were eventually agreed, Whitwick large, of 10,785 B.T.U./lb. calorific value; Whitwick 1-in. slack of 10,354 B.T.U./lb. and Cossall slack of 10,116 B.T.U./lb. The last two were similar to the slack coal already tried out on the Crosti trials. The prices of the three coals were 74/10d., 68/1d. and 57/4d. per ton compared to 88/1d. for Blidworth, the latter with a calorific value of 12,626 B.T.U.s per lb.

Dr. Giesl himself approved the engine, agreed the test procedure, and attended a number of tests. A firing instructor from the Austrian Railways, specially competent and experienced in the burning of low-grade coals with the Giesl ejector, was brought over to Rugby to advise. Thus everything was done to assure a satisfactory comparison, but, as in the case of the "Crosti" boiler, the outcome was disappointing to the protagonists.

The efficiency of the ejector as such was amply demonstrated in that maximum evaporation of 30,000 lb./in. for which the boiler resistance demanded a draught of 8 in. of water, was achieved with a blastpipe pressure of only 4½ lb./sq. in., instead of 9½ lb./sq. in. with the standard double chimney arrangement, values varying pro rata for the lower outputs. This translated itself as an example, into an increase in available indicated horsepower of from 35 to 85 i.h.p. or from 5.1 to 3.45 per cent over that obtainable at 40 m.p.h. over a range of steaming from 12,000 to 28,000 lb. per hour. Boiler efficiency remained identical in the two cases.

Reduction in blastpipe pressure does not, however, produce the dramatic reductions in steam consumption which might be imagined, and about 1¼ per cent reduction in steam per i.h.p. for each pound less of back pressure is the optimum which can be theoretically expected. Under the conditions of the agreement, namely 18,000 lb/hr. evaporation at 40 m.p.h. No. 92250 burned 1,993 lb. of Blidworth coal per i.h.p./hr., as against 2,070 lb. for the standard double blastpipe engine, a reduction of 3.72 per cent. At this rate of working, blastpipe pressure was 1.5 lb./sq. in. for the Giesl, and 3.5 lb./sq. in. for the standard engine.

Whitwick large coal was found to be too large. Even after an appreciable amount of breaking up, there were still two many voids on the grate through which the air passed without burning the coal satisfactorily, and acceptable performance was only gained after breaking it up still further into the approximate size of Blidworth cobbles. Even so consumption per i.h.p. was 2,462 lb., 20.1 per cent increase on the control conditions.

With the two slack coals the conditions of the Crosti tests repeated themselves. Both coals caused clinkering up of the front part of the grate in a random manner so that no stability of steam outout could be maintained, resulting in widely fluctuating boiler pressure and/or water levels. After every effort had been made to burn the stuff to the best advantage, consumption per i.h.p. hour was increased to 3,390 for Whitwick slack and 5,180 for Cossall slack, increases over standard of 63.8 and 150 per cent respectively. These coals were in fact just not suitable for burning within the restrictive conditions of a locomotive firebox and ashpan.

There was nothing in these results to encourage the then Transport Commission to convert any of its engines to the Giesl system, although No. 92250 retained its ejector during its remaining life. Confirming long experience, a small coal saving on an indicated basis, and under ideal stationary plant conditions, can become very much reduced or even disappear altogether in the rough and tumble of daily service, and No. 92250 did not at its home depot distinguish itself above its fellows. As regards the low-grade coal, the Coal Board could not bring itself to reduce the cost of its products to such a further extent that, notwithstanding unreliable operation, some economic advantage might have been possible by their use.

I have concluded this chapter by covering the tests with the Crosti boiler, the mechanical stoker and the Giesl ejector in some detail, in order to illustrate firstly the very thorough way in which possible improvements to the conventional locomotive were explored, at considerable cost to British Railways, and secondly to confirm how difficult it was to make appreciable further economic gains so long as the basic Stephensonian conception was retained. There are those who think that justice was not done, and that with more effort and with more faith, any or all of these improvements might have rung the bell and justified wide adoption on our lines. After reading again most carefully all the correspondence and reports, I can find little trace that such could have been the case. It was never sufficient just to have something different which somebody else had succeeded in using profitably elsewhere. It is cash saved or performance enhanced here on our home ground which speaks, and the voice was too faint in these particular cases.

Although it may have been rather tough going for the reader to take in all the facts and figures set out in such concentrated form in this chapter, I hope that certain points will have emerged sufficiently clearly Previously work on the Swindon Test Plant and associated road testing, which only sprang into real life after the last war had already laid a pattern of knowledge, particularly as regards draughting which was available during the design stage of the B.R. locomotives. As they were successively built full-scale tests on the same plant plus the new plant at Rugby, again associated with scientifically controlled road tests, explored the whole spectrum of performance and efficiency. Initial deficiencies, such as weaknesses in the draughting were spotted at once, and were rectified by simple adjustments of blastpipe and/or chimney dimensions. It was only where tests were not undertaken or where they could not be sufficiently extended, that any mysteries remained, as in the case of the "Clan" class engines which probably required very little to enhance their reputation, and as in the case of No. 71000, where we might have learned much of interest regarding the limits of combustion, but little of use, alas, for there were to be no more steam locomotives to which it could be applied.

For the rest, the family of B.R. locomotives emerged from these rigorous tests as having performance and efficiency potential, not only as good as any of the regional designs which had immediately preceeded them, much better in a few cases, but they also attained the optimum which could be reached by the simplest and most robust design embodying all those features favourable to economical building, servicing and maintenance, which we had held to be of equal importance to prowess on the line and economy in fuel. On the other hand, these results were produced under ideal conditions, they had to be if truly comparative values were to be obtained unadulterated by all kinds of extraneous and unmeasureable collateral circumstances. The engines were new built with nothing worn or deteriorated, the fuel was the best of its kind, and driving and firing was expert and closely supervised. It now remained to see how they would perform in the rough and tumble of life as it is lived out in daily service, driven and serviced by all sorts and conditions of men, in every state of deterioration, and digesting coal which could run the gamut from A to Z in quality.

VIII

Performance in Service—Classes 7 and 8

SO FAR in this book, matters within the personal knowledge of the author have been recorded, but for this chapter and the next it is necessary for him to don borrowed plumes and to draw upon the work and experience of others. No organisation exists nor has existed on British Railways or its former constituents for the systematic recording and analysis of locomotive running in daily service in the form originally introduced at the turn of the century by the late Rous Marten and greatly developed in succeeding years by C. J. Allen and O. S. Nock. The work of these gentlemen, recorded in the *Railway Magazine* and Ian Allan's periodicals, has built up through the years a body of information about engine working on the line under all the realistic conditions of weather, operating conditions, variation in fuel supplies and competence or otherwise in driving or firing which is available from no other source. For this reason, these articles have always been avidly scanned by professional railway officers, not excluding the C.M.E.s themselves, and those engaged in designing, building and operating locomotives have shared with lay enthusiasts the warm glow which comes from satisfying records of competence for the types in which they are interested, just as a continuing tale of shortfall in expectations has more than once sparked off technical investigations into the reasons therefor.

It is realised that the records in these articles are not all taken by the authors of them in person, but many in turn are the work of devoted and knowledgeable enthusiasts, for the most part amateurs in the railway engineering sense, but experts in the intricate exercise of stop-watch timing and speed assessment.

My own enthusiasms have never led me in this particular direction, and after a brief spell of attempted train timing in the first months of my start as an apprentice at Horwich Locomotive Works, when I travelled thereto each weekend from my home near Liverpool, I have never subsequently taken note of passing times and speeds. Perhaps I was dispirited by these initial experiences in which a variety of Aspinall 2–4–2 tanks and unsuperheated 4–4–0s, never loaded above 120 tons, took anything up to 28 minutes to accomplish the 18.9 miles from Liverpool Exchange to Wigan under war time conditions. In any case I preferred then and after to rely for my train running information upon the expert train timers above mentioned, and with full acknowledgements to Messrs. Allen and Nock and to their publishers past and present I have freely drawn upon their work for the contents of these chapters.

Those who have already read the descriptions herein repeated are begged to bear with the author, who has here reassembled them for the specific purpose of illustrating the points he wishes to bring out in respect of B.R. standard locomotive performance.

The volume of the records thus available is immense. The few from amongst them for which there is space in this chronicle have been picked out either to indicate how far what was designed into the engines, and demonstrated on test, emerged as actual performance potential or to illustrate some particular feature of working. The runs now to be described are thus for the most part the best of their kind, but a commentary is offered upon all the factors which could diminish this bright picture, and lead to indifferent running, factors which might be inherent in the particular locomotives themselves, but which much more often were due to the countless external circumstances to which steam locomotive operation was subject.

It is natural that the greatest number of records concern the 55 "Britannia" engines which of all the B.R. standard types, engaged most frequently in important main-line express services. Let us start with the Crewe to Carlisle section of the London Midland Region main line to Scotland where at 12.50 p.m. on January 12th, 1951, only ten days after completion, No. 70000 "Britannia" headed 442 tons of empty stock including dynamometer car on a non-stop run to Carlisle, booked 174 minutes for the 141.5 miles. The section of this run from Preston northwards is set out in the first column of Table XXI, 105 minutes being allowed at that time from passing Preston at a reduced speed of 15 m.p.h., as against a timing of 99 minutes introduced later for the "Midday Scot" train. The second column, the result of some assiduous work on the part of a gentleman writing under the pseudonym of "45671" in *Trains Illustrated* for September 1960, shows corresponding timings calculated from test bulletin data on a basis of 22,000 lb. of steam and 3,000 lb. of coal per hour, with the 420 tons trailing train load specified in the working timetable for "limited load" working over this route. It is clear from these figures that No. 70000 was on this, her first run, working not far off the capacity of a single fireman over certain parts of the route. Had the fireman in fact worked at this rate, which the schedule did not call for, the 35 minutes 5 seconds of the test run over the hardest section from Carnforth to Shap Summit would have become 30 minutes 50 seconds, but on the other hand while the same maximum steady rate would have occupied 9 minutes 50 seconds from Tebay to the Summit, No. 70000 surmounted the incline in 7 minutes 5 seconds at a minimum speed of 36 m.p.h. At this latter point cut-off was 50 per cent with full regulator and steam was being used at the rate of 35000 lb. per hour, for which under steady conditions over 5,000 lb. of coal per hour would have been required. Although for the short run up Shap, the fireman was working all out at a rate of around 4,000 lb. of coal per hour, the difference between what he was producing directly, and the greater volume of steam which

Table XXI
L.M. Region—Preston-Carlisle

		1			2			3			4			5		
	Engine / *Load Tare/Full*	*Engine 70000 "Britannia" 440 Tons (Tare)*			*Class 7 4-6-2 "Britannia" 420 Tons (Tare) Calculated Timings 22,000 lbs. water/hr.*			*Engine 70014 "Iron Duke" 375/390 Tons*			*Engine 70054 "Dornoch Firth" 433/465 Tons*			*Engine 71000 "Duke of Gloucester" 437/465 Tons*		
Distance		*Schd.*	*Time*	*Speed*	*Schd.*	*Time*	*Speed*	*Schd.*	*Time*	*Speed*	*Schd.*	*Time*	*Speed*	*Schd.*	*Time*	*Speed*
miles			*m. s.*			*m. s.*			*m. s.*			*m. s.*			*m. s.*	
0·0	Preston	—	0 00*	—	—	0 00	—	—	0 00	—	—	0 00	—	—	0 00*	—
1·3	Oxheys	3	4 05	32	3	3 43	39	—	—	—	3	5 48	32	—	—	—
4·7	Barton	—	—	60	—	—	—	—	7 57	—	—	—	—	—	7 35	—
9·5	Garstang	11	13 10	66	11	12 09	67	11	12 17	—	11	16 17	57	11	11 43	72
15·3	Bay Horse	—	—	64	—	—	66½	—	17 33	68	—	22 17	59/64	—	—	—
21·0	Lancaster	22	23 40	60	22	22 34	69/60	22	22 50	—	24	28 56	—	22	21 53	—
												0 00				
	Hest Bank			70	—	—	71	—	—	74	—	—	52	—		
27·3	Carnforth	28	29 15	70	28	28 13	67/71	28	28 10	—	8	9 15	56/61	—	27 10	72
34·6	Milnthorpe	—	—	64	—	35 45	66/62½	—	34 59	65	—	17 17	59½	—	33 45	68
40·1	Oxenholme	43	41 55	52	43	41 40	44½	43	42 10	—	23	23 59	50/44	—	39 32	54
47·2	Grayrigg	—	—	40		53 10	34	—	54 58	28	—	34 03	39	—	48 35	41
49·0	Low Gill	—	—	—	—	—	55½	—	—	—	—	36 04	51	—	†	
53·2	Tebay	60	57 15	64	60	59 13	61½	60	61 34	66	40	41 10	45/52	—	55 15	25
56·2	Scout Green	—	—	—	—	—	30½		65 22			45 15	33	—	—	—
58·7	Shap Summit	70	64 20	36	70	69 03	24	70	71 09	25	50	51 05	23	—	67 06	28
60·8	Shap	—	—	62	—	—	53½	—	—	—		54 14	56	—	—	—
72·3	Penrith	83	80 00†	56	83	82 43	60/52	83	84 02	—	63	64 53	63	—	79 23	—
77·4	Plumpton	88	84 15	74	88	87 55	65/61½	—	—	—	68	69 41	68/67	—	—	—
83·2	Southwaite	—	—	75	—	—	70	—	—	77	—	74 39	73	—	—	—
90·1	Carlisle	105	97 55	—	101	100 58	—	101	100 35		81	81 54	—	99	96 30	—

**Pass †Slack for P.W.S. or Signals*

the engine was consuming was made up by drawing upon the large reservoir or heat storage capacity of the boiler. This generous capacity was emphasised by the fact that the boiler pressure was still 240 lb./sq. in. at the Summit, with 225 lb. at the steam chest, although boiler water level was naturally falling. Test results confirm that on this run, maximum indicated h.p. was 2,100, and drawbar h.p. referred to constant speed on the level was 1,870, very nearly the utmost of which the engine was capable. The remainder of the run down to Carlisle was marred by a prolonged slack down to 20 mp.h. approaching Penrith, and 77 m.p.h. was the highest speed attained before finally shutting off. As a reward for my efforts, such as they were, in the design of this engine, Riddles let me have "first go" at riding the footplate during the northward trip, and it was naturally one of the most thrilling experiences of my life, for added to the wonderfully satisfying "feel" of any steam locomotive functioning perfectly, there was in this case the added satisfaction which arose from a certain measure of creation.

The return trip next day gained 6 minutes 45 seconds on the 55 minutes booking from Carlisle to Shap Summit but otherwise strictly observed the easy 190 minutes timing to Crewe. That the engine had not exceeded the bounds of economy notwithstanding its "all out" effort up Shap the previous day, was evidenced by an overall coal consumption for the harder outward trip of only 38 lb. of coal per mile, 3.26 lb. drawbar horsepower hour.

In every series of ostensibly identical locomotives, there are always a select few which seem able to give just that little more by way of performance than do their stable mates, due probably to every possible mechanical variable which affects their working adding up on the plus rather than the minus side. Of such was No. 70014 *Iron Duke* at one time employed on the Continental Boat Expresses on the S.R., but later drafted to the L.M. Region. Column 3 in Table XXI, reported by Nock, displays this engine in a very business-like start with 390 tons gross out of Preston, her driver almost making the well nigh impractically tight timing of 22 minutes through Lancaster. Although the ascents of both Grayrigg and Shap were inclined to be leisurely, a minute of arrears on passing the Summit was recovered by a brisk and undelayed descent, and Carlisle was reached on time. Comparison with Column 2 shows how very closely this engine was being worked in ordinary traffic to a uniform steam rate of 22,000 lb./hr., a rate which assured maximum efficiency and economy.

Column 4, recorded by "45671" in the same article referred to above, illustrates some of the vagaries which could beset "run of the mill" performances. No. 70054 *Dornoch Firth* had a 4 minutes late start made in a thin drizzle of rain with the brakes dragging. Initial slipping was continuous up to No. 5 Box, and thereafter the driver indulged in the very common fault of linking up too early so that 60 m.p.h. was not attained until Bay Horse 15 miles out and 5 minutes was lost to Lancaster, where a stop was scheduled with this particular train. Up through Oxenholme to

Grayrigg the climb was well up to standard, the engine making as the record remarks, a lovely staccato bark as it forged up the hillsides towards the top. However on the falling grades from Low Gill to Dillicar troughs, the driver shut off steam, judging that he had to get more water into the boiler, with the result that Shap was attacked at no higher initial speed than 52 mp.h. Considering that there was half a glass of water with the regulator shut this was excessive caution, particularly since the water was only an inch below the top of the glass on the level when the Summit Box was eventually passed at only 23 m.p.h. Although already 10 minutes behind schedule, no effort was made to regain any time on the descent to Carlisle which would easily have been within the capacity of the engine. In fact 1 minute was actually lost from Summit down to Penrith.

It is interesting to compare the method of working this engine, in average condition, although clearly tight at the valves, with 70000 when in mint condition new out of Crewe Works. Table XXII compares the footplate observations which were taken, and it will be seen that 70054 lagged behind 70000 throughout in boiler pressure calling for increased

Table XXII

Footplate Observations—Preston–Carlisle

	Date	*Jan. 1951*				*Sept. 1958*		
	Engine *Load*	*70,000* *Britannia* *442 Tons Tare*				*70054* *Dornoch Firth* *433 Tons Tare* *465 Tons Gross*		
Distance Miles	*Observations*	*Boiler Pressure lbs/sq.in.*	*Reg.*	*Cut off %*	*Steam Temp. °F*	*Boiler Pressure lbs/sq.in.*	*Reg.*	*Cut off %*
0·0	Preston	250	Full	40/25	570	240	½	70/42
1·3	Oxheys	240	,,	25	629	245	Full	33/25
7·4	Brock	245	,,	20	659	210	,,	25
9·4	Garstang	245	½	20	659	230	,,	25
15·2	Bay Horse	—	½	—	—	225	,,	—
21·0	Lancaster	235	½	20	629	235	,,	40/33/25
—	Hest Bank	240	½	20	639	225	,,	25
27·3	Carnforth	245	½	20	639	220	,,	25
—	Burton and Holme	245	¾	20	659	225	,,	25
34·6	Milnthorpe	250	¾	20	659	215	,,	30/34
40·1	Oxenholme	240	Full	30/33	668	240	,,	38
—	Lambrigg	230	,,	40	688	230	,,	38
47·2	Grayrigg	230	,,	40	688	225	,,	25/22
49·0	Low Gill	240	,,	27	678	220	Shut	—
53·2	Tebay	245	,,	37/43	698	235	Full	30/35
56·2	Scout Green	250	,,	45	713	215	,,	40/45
58·7	Shap Summit	240	,,	50	718	210	,,	45
60·8	Shap	200	Shut	30	708	200	,,	25/13
—	Thrimby Grange	—	—	—	—	—	Shut	—
72·3	Penrith	200	—	30	639	—	¾	16
77·4	Plumpton	200	Shut	30	601	225	¾	13
—	Calthwaite	—	—	30	—	225	¾	—
83·2	Southwaite	—	—	—	—	210	Shut	—
90·1	Carlisle	250	—	—	—	—	—	—

cut-offs on the level sections. There is some evidence that the safety valves were blowing off light and it could easily have been of course that the tube ends were birds nested and were not entirely clean as the engine left Manchester. The quality of the coal also may have been a factor. Apart from this, and in spite of copy book working on full regulator wherever possible here was a performance whose very ordinary nature was also due to certain ingrained habits on the part of a senior driver on the point of retirement. Premature notching up which so delayed the start was a driving habit extraordinarily difficult to instruct or argue a man out of, if that was his natural impulse. The shutting off of steam beyond Low Gill was pure misjudgment; also the refusal to run and regain time downhill from Shap, was another individual manifestation not necessarily connected with any rough riding at the back end at the highest speeds.

We now turn to three runs over another route selected to show the very marked ability to run fast which the Britannias possessed, a feature not wholly brought out in the official test results. Although not widely allocated to Kentish Town and Trafford Park depots on the Midland division of the L.M.R. until 1958, the drivers and firemen of those sheds were able to coax the fastest running out of these engines which was experienced anywhere, and the numerous published records of their performances on this line are freely studded with "90s". To do this the engines must often have been in better condition than they looked, for the days of Wetherburn and Talbot at Kentish Town were far behind when an immaculate turn-out was *de rigueur*, and most of them, which I frequently saw in those days at St. Pancras or passing Radlett, were of a distinctly scruffy appearance.

The ubiquitous *Iron Duke*, No. 70014, appears in the first run of Table XXIII on the down "Palatine" express over the 99 miles to Leicester with booked stops at Luton and Wellingbrough. Nock reports that a speed of 98 m.p.h. was recorded at Flitwick on the downhill run towards Bedford, but that thereafter speed was eased back to 80-83 m.p.h. on what is normally the fastest part of the descent. Having reached almost level track, however, the engine was opened out again; Bedford was passed at 90 m.p.h., speed was held at 82 to 88 right on to Sharnbrook Station and the three-mile bank at 1 in 119 was cleared at the astonishing minimum speed of 73½ m.p.h. By the summit at Milepost 59¾ the train was again on time after a 3½-minute late start from Luton due to previous checks. Yet another "90" was recorded at Irchester, but the train was heavily checked between Kettering and Desborough North, thereby losing another 5 minutes on schedule. Again there was a thrilling recovery in that after slowing carefully for the Market Harborough curve speed was worked up to 73½ m.p.h. at East Langton, but then, up the 3½ miles to Kibworth North, mostly steeper than 1 in 200 and including two miles at 1 in 130-114, the speed never fell below 70 m.p.h. This was followed by a swift acceleration to yet another "90" before the Wigston slack. The net times over the three sections of this run were 31, 28¼ and 34 minutes respectively to compare with the booked timings shown in the table, and indicate something of the capacity of these engines for handling these tightly timed short distance runs.

The second run in Table XXII was reported by Allen in his book *British Pacific Locomotives*. Here No. 70032 *Tennyson* and a nine-coach train of 320 tons, lost over 5 minutes in the early stages of a through run due to signal and permanent way delays, but then followed a performance so startling in character that between St. Albans and Leicester over 8½ minutes was gained on schedule. After climbing the 4 miles at 1 in 176 to Sandridge 1.4 miles beyond St. Abans at a minimum of 65 m.p.h., some very high speed running began, culminating in a top speed of 99 m.p.h. at the foot of 5½ miles down 1 in 202 to Flitwick; the average was all but 90 m.p.h. for 17.2 miles from Leagrave to Bedford, and 93 m.p.h. over the 7.75 miles from Harlington to milepost 45. The 3¾-mile climb to Sharnbrook summit was breasted at 57½ m.p.h., and with the train now on time there was a slight easing, although the running was still very fast. Eventually Leicester 99.1 miles was reached in 95 minutes 40 seconds, but

Table XXIII
L.M. Region—St. Pancras-Leicester

		1			2		
Distance Miles	*Engine* / *Load Tare/Full*	*70014 Iron Duke 282/300 tons*			*70032 Tennyson 297/329 tons*		
		Schd.	*Time m. s.*	*Speed*	*Schd.*	*Time m. s.*	*Speed*
0·0	St. Pancras	0	0 00	—	0	0 00	—
1·5	Kentish Town	4	4 25	—	4	6 42	—
6·9	Hendon	10½	14 42	—	10½	13 56	52½
12·4	Elstree	—	20 12	—	—	21 51	52½
15·2	Radlett	—	22 30	—	—	24 22	76½
19·9	St. Albans	23	26 05	—	23	28 12	65
24·6	Harpenden	—	32 27	—	—	32 23	73½
30·2	Luton	34	37 10	—	33	36 46	81½
32·7	Leagrave	—	4 11	—	—	38 38	78½
37·2	Harlington	—	8 00	—	—	41 42	94
40·2	Flitwick	—	—	98	—	43 32	99
41·7	Ampthill	—	11 00	83	—	44 30	94
49·9	Bedford N.	18	17 00	90	49	50 08	77
56·7	Sharnbrook	—	21 42	—	—	55 33	70
59·7	MP. 59¾	27½	24 10	73½	58½	58 32	57½
62·7	Irchester	—	26 12	90	—	60 51	83½
65·0	Wellingboro'	34	28 21	—	63½	62 40	62
68·2	Finedon	—	5 25	—	—	65 32	71½
72·0	Kettering	8½	8 48	—	70	68 48	66
78·5	Desboro'	—	20 30	—	—	75 31	—
82·9	Market Harboro'	20½	25 15	—	82	79 35	—
86·3	East Langton	—	28 28	73½	—	82 33	72
89·7	Kibworth	—	30 38	70	—	85 38	61½
91·5	Great Glen	—	32 42	90	—	87 18	75
95·4	Wigston	—	35 27	—	—	90 31	—
99·1	Leicester	38	40 06	—	99	95 40	—
	Net Time					89½	

the net time of 89½ minutes was 9½ minutes less than the 99 minutes allowed.

The first run in Table XXIV, this time in the up direction, was published in 1960. This was with ten coaches, and Allen, who timed the run, reported that seldom if ever had he known anything on the Midland to equal the running from Leicester southwards, especially in view of the 355-ton load. The start from Leicester was 4 minutes late, and the running through Market Harborough was on the lively side; it was followed by a minimum of exactly 60 m.p.h. up the 4 miles at 1 in 132 to Desborough North. Beyond Kettering 90 m.p.h. was reached, and then followed another rousing climb in which the 4 miles to Sharnbrook Summit—the last 2½ at 1 in 120—was surmounted at a minimum of 61 m.p.h. From Market Harborough to Bedford alone there had been a gain of 5½ minutes on this "XL" schedule. Up the long climb thereafter to milepost 34, however, there was not only a permanent way delay at

Table XXIV

L.M. Region—Leicester-St. Pancras

		1			2		
Distance Miles	*Engine* / *Load Tare/Full*	*70014 Iron Duke 336/355 tons*			*73073* *302/330 tons*		
		Schd.	*Time m. s.*	*Speed*	*Schd.*	*Time m. s.*	*Speed*
0·0	Leicester	—	0 00	—	—	0 00	—
3·7	Wigston	—	6 59	—	—	6 32	—
7·5	Great Glen	—	10 55	—	—	11 16	—
9·4	Kibworth	—	12 55	56½	—	13 56	—
12·8	East Langton	—	15 46	83½	—	16 18	—
16·2	Market Harboro'	19	18 19*	—	19	19 14	—
20·6	Desboro'	—	22 34	60	—	24 18	—
27·1	Kettering	31	27 34	90	31	29 49	—
30·9	Finedon	—	—	—	—	32 55	—
34·1	Wellingboro'	37	32 33*	—	37	35 38	—
—	Irchester	—	34 32	—	—	—	—
39·4	MP 59¾	43	37 17	61	43	40 31	—
42·4	Sharnbrook	—	39 41	92	—	43 07	83
49·2	Bedford N.	51	44 43	80	51	48 28	—
57·3	Ampthill	—	56 59†	—	—	55 33	—
—	Flitwick	—	— ‡	—	—	—	—
61·8	Harlington	—	64 48	—	—	59 34	—
66·3	Leagrave	—	69 58	52	—	63 51	61
68·9	Luton	70	72 11	75	70	65 59	—
74·5	Harpenden	—	76 30	76½	—	70 26	—
79·2	St. Albans	79	80 01	84	79	74 09	—
83·9	Radlett	—	83 12	92	—	77 42	—
86·7	Elstree	—	85 11	79	—	79 56	—
92·2	Hendon	90	90 31‡	72	90	84 19	—
97·6	Kentish Town	96	96 59‡	—	96	89 16	—
99·1	St. Pancras	99	102 06‡	—	99	92 05	—
	Net Time	88			92		

**Speed restriction* †*Signal stop* ‡*Slack, PWS or Signal*

Wilshamstead but also signal checks from the 1.57 p.m from Sheffield running late in front. This turned a 6¼-minute gain on schedule into a loss of 2¼ minutes by Luton. Now came another terrific effort which cut all Allen's previous records over the 17.8 miles from Luton to Elstree down to 13 minutes precisely, with a top speed of 92 m.p.h. at Radlett and a minimum of 79 m.p.h. over Elstree summit. In spite of this energy, permanent way slowings at Mill Hill and Kentish Town and a final check to walking pace outside St. Pancras led to a 7-minute late arrival. With due allowance for the slacks, net time for the run was 88 minutes. Need it be added that the engine was No. 70014 *Iron Duke*?

By far the greater number of Britannia runs which have been published have been on the Great Eastern section of the Eastern Region, the area to which these engines were first sent, and for which, unlike their other allocations, they were something far ahead in capacity over what had previously been used. Moreover from their first arrival until their final removal from Stratford and Norwich sheds 11 years later in face of the advancing diesel, they were lovingly looked after, and kept in beautifully clean condition whilst those in charge of the depots and the men who drove and fired them were enthusiasts all. In such an atmosphere it is

Table XXV

Eastern Region—Ipswich-Norwich

		1			*2*		
Distance Miles	*Engine* *Load Tare/Full*	*70035 Rudyard Kipling 310/325 tons*			*70007 Coeur de Lion 304/320 tons*		
		Schd.	*Time m. s.*	*Speed*	*Schd.*	*Time m. s.*	*Speed*
0·0	Ipswich	—	0 00	—	—	0 00	—
0·8	E. Suffolk Jn.	—	2 09	—	—	—	—
2·4	Bramford	—	3 51	63½	—	4 23	—
4·8	Claydon	—	6 02	68	—	6 42	—
8·4	Needham	—	8 57	75½	—	9 48	—
11·9	Stowmarket	—	11 39	82	—	12 48	72
14·2	Haughley	14½	13 25	74½	—	14 46	—
16·7	Milepost 85½	—	15 35	70½	—	17 06	—
17·9	Finningham	—	16 29	77	—	18 02	—
22·7	Mellis	—	19 58	87½	—	21 40	—
26·3	Diss	—	22 21	94	—	24 06	93
28·8	Burston	—	24 02	86½	—	25 48	—
31·9	Tivetshall	29	26 12	81	—	28 07	79
35·4	Forncett	—	28 50	83½	—	30 40	—
38·0	Flordon	—	30 38	92	—	32 25	—
41·0	Swainsthorpe	—	32 46	86½	—	34 29	—
43·2	Milepost 112	—	34 25	81	—	36 12	—
44·2	Trowse Upper Jn.	—	35 19‡	37	—	37 20	—
45·3	Trowse	—	37 03	—	—	39 08	—
46·3	Norwich	44	39 07	—	—	41 13	—
	Net Time					41¼	

‡*Slack, PWS or Signal*

small wonder that they maintained a high level of performance throughout the whole period on the Cambridge, Norwich and Parkstone Quay runs. Out of a wealth of records I will confine myself for space reasons to running on the Ipswich–Norwich part of the Great Eastern main line. Column 1, Table XXV, recorded by Allen is probably one of the fastest runs ever made over Great Eastern metals. No. 70035 with a nine-coach train of 325 tons gross was late away from Ipswich on one of the 44-minute bookings. On the second of the level stretches on the gradual rise up the Gipping Valley, the 4–6–2 reached the unusual maximum of 82 m.p.h. at Stowmarket and this was sufficient to carry the express up Haughley bank—2 miles at 1 in 131—at a minimum of just over 70 m.p.h. Then 26½ miles of the undulating stretch that follows, between milepost 85½ (Finningham Summit) and milepost 112 were reeled off at an average of 84.4 m.p.h. with a full mile past Diss at 94 m.p.h., so that milepost 112, 43,2 miles from the start was passed in 34 minutes 25 seconds. Notwithstanding nearly 4 minutes spent over the last 2 miles, the stop was effected at Norwich Thorpe in 39 minutes 7 seconds, a start to stop average of 71 m.p.h.

Column 2 was representative of best daily efforts without the spur of a late start and was also published by Allen in 1952. In this case the

Table XXVI

Eastern Region—Norwich-Ipswich

		1			2		
Distance Miles	*Engine* / *Load Tare/Full*	*70000 Britannia 263/280 tons*			*70000 Britannia 274/295 tons*		
		Schd.	*Time m. s.*	*Speed*	*Schd.*	*Time m. s.*	*Speed*
0·0	Norwich	—	0 00	—	—	0 00	—
—	Trowse	—	—	—	—	2 56	—
—	Trowse Upper Jn.	—	—	—	—	6 00	—
—	Milepost 112	—	—	—	—	—	—
5·3	Swainsthorpe	—	8 54	55½	—	10 08	63
8·3	Flordon	—	11 58	68	—	12 52	79½
10·9	Forncett	—	14 13	70	—	14 50	81
14·4	Tivetshall	18	17 11	68	17½	17 30	78
17·5	Burston	—	19 43	75/72	—	—	—
20·0	Diss	—	21 42	78	—	21 22	91½
23·6	Mellis	—	24 45	69	—	23 54	79
28·4	Finningham	—	28 56	76	—	27 30	81
—	Milepost 85½	—	—	—	—	28 18	78
32·0	Haughley	35	31 48	78	32	30 15‡	75
34·4	Stowmarket	37	33 44‡	71½	—	31 58	82½
37·9	Needham	—	37 57	61	—	34 28	82
41·4	Claydon	—	41 18	66	—	37 10	78
43·8	Bramford	—	42 32	64½	—	39 00	76
—	E. Suffolk Jn.	—	—	—	—	40 30‡	—
46·3	Ipswich	51	47 04	—	45	42 16	—
	Net Time		45½				

‡*Slack, PWS or Signals*

"Broadsman", allowed 45 minutes for the run, was the train concerned with No. 70007 in front. Speed was up to 72 m.p.h. by Stowmarket; the minimum speed up Haughley bank, although not recorded was probably about 62 or 63 m.p.h. Then came a lightning swoop down to Diss with a top speed of 93 m.p.h. while at Tivetshall no lower rate than 79 m.p.h. was recorded. The 5.6 miles from Forncett to Swainsthorpe were then run at an average of 88.8 m.p.h. so helping towards a mean speed of 83.9 m.p.h. over the 25.4 miles from Finningham to milepost 112, and to the overall time of 41 minutes 13 seconds.

In the opposite direction, Column 1, Table XXVI, has interest as the first published run in May 1951 with "Britannia" after allocation to her home station at Stratford. Although the men were still feeling their way with the new engine, and had not yet established the confidence which supported later more brilliant performances, none the less the engine was still at this early stage beginning to show its paces. On a timing of 51 minutes northbound as it was then, No. 70000 made a gentle start but after Forncett was given her head, and as Allen reports, subsequent up gradients might not have been there for any notice the engine took of them. Up 4½ miles of mostly rising grades past Forncett including two separate 1¼ mile stretches at 1 in 136 and 1 in 156, speed was a steady 68-70 m.p.h.; down past Burston a quickly-attained 75 was moderated to 72 but soon rose to 78 after Diss; up 1½ miles at 1 in 132 to Mellis the minimum was 69 m.p.h. but the rest of the climb to milepost 85¼ including 2¼ miles at 1 in 248 was surmounted at 72 m.p.h. After that the engine could have attained unlimited speed down the Gipping Valley, but was not allowed to exceed 78 past Haughley, was slowed to 71½ m.p.h. through Stowmarket, and then more severely for a p.w. check to 35 m.p.h. which cost about 1¾ minutes. So the actual time for the 46.3 miles with its awkward start out of Norwich was 47 minutes 4 seconds and the net 45½ minutes, rather better than "even time". What the same engine could do when it settled down is shown in Column 4, recorded some nine years later in 1960. Speed did not fall below 78 m.p.h. up the rising grades to Tivetshall nor below 79 m.p.h. up 1¾ miles at 1 in 132 to Mellis, while the final rise to milepost 85½ was carried by storm at 78 m.p.h. The 14.35 miles from Diss over Finningham Summit to Stowmarket were reeled off at an average of 81.5 m.p.h.

Allen adds in the same article which records the above that the fastest Britannia exploit he had ever known up to that time south bound from Norwich was with No. 70003 *John Bunyan* with a nine-coach load of 325 tons gross when 27.05 miles from Tivetshall to Claydon, between checks, were reeled off in 18 minutes 57 seconds at an average of 87.5 m.p.h. Speed before Diss was 92 m.p.h.; 82 was the minimum at Mellis and also the lowest rate on the ascent to Finningham.

These samples are only a fragment of the large score of times runs which these engines logged up, showing a remarkably high standard of consistency and good running over a none too easy line where they were looked after well and driven with appreciation.

In the latter connection and in the very act of completing this chapter word has come to me of an exploit on almost the last day before the withdrawal of the Britannias from the G.E. section, which confirms that they had lost no whit of their sparkle, after ten years of arduous service. It is contained in a letter from Richard Hardy who was in charge of Stratford District at the time, and I cannot do better than quote his own words:—

> "When the diesels began to come, the remaining Britannias went to Norwich and we used them. They got the best of maintenance and how we used to run them! . . . the semi-fasts had a lot of time knocked off them and trains like the 1.45 ex Norwich had to be by Tivetshall in 16 minutes, as against the original Britannia timing of 18 minutes. I always said I could start Brentwood bank at 60 and finish it at 60 with nine coaches. I never got the chance to do just that until the Friday before the end of the Britannias on the G.E. With the Essex Coast Express I took 70009 down, kept to the 60/62 m.p.h. restriction all the way to the foot of the bank, and let her out gradually from 15-50 per cent cut-off and a shade over. We topped the bank at 61 on the speedometer, with 250 lb. of steam and nearly a full glass of water in the boiler."

Such keenness from the top naturally inspired all concerned, and it was only with such keenness that the best could be got out of any class of steam engine. If the engine responded to such attention with vivid performance, economical working and reliable service, then the satisfaction was mutual, and I think it is only just to mention the names of those in charge of the Britannias during their life on the G.E. section when they certainly rose to the heights of which steam traction was capable. Before Hardy himself, T. C. B. Miller, now Assistant General Manager L.M. Region, was in charge at Stratford when the Class 7s first arrived, while R. L. Vereker ruled at Norwich followed later by D. Harvey who saw the engines out at that depot.

The relative absence of a comparable number of logs on the Western Region covering identical engines of the same class can only reflect the very different conditions of acceptance by the footplate men and many of their superiors on that region, the reasons for which will be discussed later. On my own footplate runs, which ranged at different times from Paddington to Penzance and to Newport, I was conscious of no falling off of expected performance, but clearly the presence of the designer of the engines accompanied by a senior inspector had its inevitable psychological effect of assuring intelligent driving and firing. No more than this was required to enable the engines to deploy all the good that was in them. Be this as it may, hardly any logs of distinction have been published over the Plymouth main line and it was only after all the engines allocated to the Western Region had been concentrated at Cardiff and subjected to understanding supervision that really good running was at last forthcoming. A fair example of what the engines were able to do on this route is included here to round off this bird's eye view of Britannia capabilities.

It is taken from Allen's book, *British Pacific Locomotives*, and is set out again in Table XXVII. No. 70028 *Royal Star* had 13 bogies, 445 tons tare, 475 tons gross to take from Newport to Paddington in 2½ hours of continuous steaming. The overall timings were not at that time very exacting, and in particular had easy allowances through the Severn Tunnel to permit a margin in the event of freight train delays. "Royal Star" took full advantage of this latter circumstance and up the 3¼ miles at 1 in 100 from the centre of the tunnel to Pilning speed dropped from 73 to 36 m.p.h., and after a momentary recovery at Cattybrook, from 40 to 20 m.p.h. up the ensuing 3¼ miles at the same grade to Patchway.

Table XXVII
Western Region—Newport-Paddington

Distance	Engine Load Tare/Full	70028 "Royal Star" 445/475 Tons		
		Sched.	*Time*	*Speed*
miles			*m. s.*	
0·0	Newport	—	0 00	—
3·7	Llanwern	—	6 47	54
7·4	Magor	—	10 25	64
9·8	Severn Tunnel Jn.	13	13 18	30
10·7	Severn Tunnel West	14½	14 59	52/73
15·6	Severn Tunnel East	—	19 46	40
16·7	Pilning	—	21 39	36
18·1	Catty Brook	—	24 01	40
20·2	Patchway	31½	29 24	20½
21·5	Stoke Gifford West	33½	31 26	38
24·9	Coalpit Heath	—	36 01	49
28·8	Chipping Sodbury	—	42 36	—
			43 27	
33·4	Badminton	49½	51 45	44
39·1	Hullavington	—	58 07	63
43·6	Little Somerford	—	62 08	72
46·4	Brinkworth	—	64 32	65
50·5	Wootton Bassett	67½	68 51	53
56·1	Swindon	73	74 55	59
61·9	Shrivenham	—	80 29	64
66·9	Uffington	—	85 05	67
69·6	Challow	—	87 33	67
73·0	Wantage Road	—	90 39	68
76·9	Steventon	94	94 03	70
80·3	Didcot	97	96 54	71
84·9	Cholsey	—	100 55	70
88·6	Goring	—	104 04	72
91·8	Pangbourne	—	106 44	73
94·7	Tilehurst	—	109 07	72
97·4	Reading	113	111 26	67
102·4	Twyford	117½	115 52	69/25
109·1	Maidenhead	123½	123 15*	—
114·9	Slough	129	130 26	66
124·3	Southall	138	138 30	72
127·7	Ealing Broadway	—	141 29	66
133·4	Paddington	149	149 25	—
	Net Time		140	

***Signal Slack**

Speed had recovered to a steady 49 m.p.h. up the long 1 in 300 from Stoke Gifford when there came a dead stand for a minute at Chipping Sodbury due to a preceding freight train not having cleared into Badminton loop.

From the re-start 44 m.p.h. was attained up the remainder of the 1 in 300 to Badminton. A higher speed might have been expected at Little Somerford than 72 m.p.h., but no doubt the boiler was being refilled after the hard climb. Onwards from Swindon, while the speed down the 1 in 754 to Didcot was mostly kept at 67-68 m.p.h., on the level from Didcot it rose to a steady 70-73 m.p.h., so that by Reading the train was 1½ minutes early. After the Maidenhead signal check there was another 72 m.p.h. at Southall, and with a cautious approach, Paddington was reached on schedule in 149 minutes 25 seconds. The net time of 140 minutes showed a gain of 9 minutes. From Chipping Sodbury the 104.55 miles to London had been run in 105 minutes 58 seconds start to stop or 101½ minutes net; from Hullavington to Ealing Broadway the time over the 88.55 miles pass to pass was 83 minutes 22 seconds or just under 79 minutes net. Inspection of the figures in this table, referred to the test bulletin for this class of engine indicates that here again a steam consumption of 22,000 lb. per hour and a firing rate of about 2,700 lb. of coal per hour was being maintained over the greater part of the journey, at which rates the engine was working at very nearly its optimum efficiency.

As a recorder of so much of the work of these engines in all circumstances over the various routes upon which they ran, Allen sums up: "It is a pity that the "Britannia" Pacifics appeared on the scene so late in British Steam Locomotive history. With reasonably good maintenance and in competent hands they might have established a notable reputation. In any event the runs just described leave no doubt as to their outstanding capabilities." It is not easy to disagree with this objective assessment.

The Class 8 Pacific No. 71000 *Duke of Gloucester* was stationed at Crewe North shed and ran both north and south of that centre, its most regular working being on the "Midday Scot", whose timing of 80 minutes for the 82 miles of the Euston-Rugby portion, with loads reaching 500 tons, was one of the hardest on the London Midland region. In the course of working out the results of tests with this engine, Ell tabulated and included in Bulletin No. 15 the optimum performance which could be expected with different loads and timings over the whole route from Euston to Carlisle, so that it is possible to set this against the few recorded actual runs with this engine, and to see how closely actual and theoretical performance could correspond.

Column 5 in Table XXI describes a run to Carlisle reported by Nock in 1961. The recovery from the Preston passing slack was on the slow side, and the engine was eased considerably after a maximum speed of 72 m.p.h. had been attained at Brock troughs. But after Lancaster some excellent work was done, with a minimum speed of 41 at Grayrigg. The permanent way restriction checked speed down to 25 m.p.h. before Tebay but up Shap the engine recovered in fine style to a sustained 28 m.p.h.

From the bulletin charts it is probable that an equivalent 1,550 d.b.h.p. was being developed at a steam rate of 26,000 lb./hr. and a coal rate which according to the quality of fuel on the tender that day and to the extent to which the boiler reservoir capacity was drawn upon might have been anything from 3,000 to 4,170 lb. per hour. Table XXVIII records two runs from Euston to Rugby on the 80-minute timings against which some calculated times are shown in Column 2 derived from the test bulletin. That in Column 1 was timed by R. I. Nelson who rode the footplate, and this run has been reported both by Allen and by Nock. Full regulator working was the order of the day, save only for a few moments after each start. Out of Euston 75 per cent cut-off and first valve were used to get the train on the move with the usual rear end assistance; almost at once the regulator was opened to full as cut-off was brought back to 54, and then in stages to 40 and 27 at milepost 1, where speed at the top of Camden bank had risen to 24 m.p.h. Cut-off was next reduced very gradually until it was down to 10 per cent at Brent junction, speed here having risen to 67½ m.p.h. From Wembley a gradual increase to 13 per cent by Hatch End could not prevent a fall in speed to 56 m.p.h. up the 1 in 330, corresponding to 22,000 lb. of steam per hour. With 10 to 12 per cent

Table XXVIII

L.M. Region—Euston-Rugby

Distance Miles	Engine Load Tare/Full	1 71000 Duke of Gloucester 435/485 Tons			2 Class 8 4-6-2 480 Tons Tare Calculated Timings 26000 lb Steam/Hr.	3 71000 Duke of Gloucester 468/510 Tons		
		Schd.	Time m. s.	Speed	Time m. s.	Schd.	Time m. s.	Speed
0·0	Euston	0	0 00	—	0 00	—	0 00	—
1·0	M.P.1	—	3 16	24	—	—	3 27	20
5·4	Willesden	9	9 19	58½	8 48	9	9 45	55½
8·1	Wembley	—	11 54	67½	—	—	12 28	58
11·4	Harrow	—	15 16	58	—	—	16 04	55½
13·3	Hatch End	—	17 18	56	—	—	18 07	55½
17·4	Watford	21	21 31	64	21 12	21	22 31	64
20·9	Kings Langley	—	24 39	66½	—	—	25 52	64/30
24·5	Hemel Hempstead	—	28 33	59½/15	—	—	31 15*	50
28·0	Berkhamsted	—	34 09*	51	—	—	35 10	60
31·7	Tring	35	38 14	59	36 06	35	38 52	61
36·1	Cheddington	—	42 07	78	—	—	42 46	76
40·2	Leighton Buzzard	—	45 20	74	—	—	45 58	83½
46·7	Bletchley	47	50 27	78	48 00	47	50 55	80
52·4	Wolverton	—	54 51	83	—	—	55 08	86
54·8	Castlethorpe	—	56 44	74	—	—	57 00	74
59·9	Roade	58	61 17	63	59 12	58	61 29	67
62·8	Blisworth	61	63 50	78½/50	62 00	61	64 00	76
69·7	Weedon	67	70 21*	70½	67 48	67	69 13	85½
75·3	Welton	—	75 23	63	—	—	73 54	67½
80·3	Hillmorton	—	79 50	77	—	—	78 20	76
82·6	Rugby	80	83 47*	22	80 00	80	81 42*	18
	Net Time Mins		78				78½	

***Slack, PWS or Signals**

on the slight fall and level past Watford to Kings Langley came an acceleration to 66½ m.p.h., before the bad permanent way slowing at Hemel Hempstead; in recovering from this 31, 25 and 18 per cent cut-offs were used in succession which resulted in an acceleration to 59 m.p.h. up the second 1 in 330 stage.

Then on 7½ and 8 per cent cut-off, the engine quickly worked up to 78 m.p.h., and with a brief 10-11 per cent from Leighton up the slight rise through Bletchley to Denbigh Hall, followed by reversion to 7-8 per cent, the top speed of the journey, 83 m.p.h., was attained through Wolverton. From there onwards on cut-offs varying between 10 and 15 per cent., No. 71000 was doing no more than keep point to point times, and so arrived at Rugby 3¾ minutes late by the clock, although the checks had more than accounted for the arrears and the net time was no more than 78 minutes. It was observed that steam chest pressure with full regulator averaged only 10 lb. less than the boiler pressure; the latter fluctuated from 250 lb. at the start through 230-240 lb. most of the way to Watford, after which there was a drop to 195 lb. at the Hemel Hempstead slowing though by Tring 250 lb. had been re-attained. But at no stage did there appear to the footplate observer to be any trouble with steaming.

It will be seen from Column 2 that the working throughout was very close to that calculated from test results at 26,000 lb. of steam per hour. Although through Harrow and Hatch End the steaming rate was undoubtedly held below this figure, the effect upon the timing from Willesden to Watford was only a debit of 24 seconds.

The second run in Column 3 was reported by Allen in 1956. The load was one coach more, but although a considerable period in time separated the two runs, the one was almost a carbon copy of the other. The same caution in raising the firing and steam rates up to Watford was apparent and 1½ minutes was dropped to the latter point. The acceleration from 30 to 61 m.p.h. up to Tring was good, and then followed some high speed with maxima of 86 and 85½ m.p.h. at Wolverton and Weedon respectively, minima of 67 and 67½ at Roade and Kilsby Tunnel and an average of 74.6 m.p.h. over the 44.2 miles from Cheddington to Hillmorton. So, after passing Tring 4 minutes late, Rugby was reached no more than 1¾ minutes late, and that despite a final signal check, the net time being 78½ mins. Again it was reported that the engine steamed beautifully.

Notwithstanding the tests results recorded in Chapter VII one might well ask what could be really wrong with an engine capable of such performances, yet the flaw was deep seated, and was quickly detected by the enginemen, even if they were naturally unable themselves to put their fingers on the cause. In the first place above rates of working which might vary from 20,000 to 26,000 lb. of steam per hour according to quality of coal, original condition of the fire and above all the skill and attitude of the crew, serious deterioration in the condition of the firebed could set in due to the effects of vibration and other causes. Secondly even before the point at which combustion conditions went awry, boiler efficiency was already falling to an extent which more than counterbalanced the very economical use of steam in the cylinders. Thus when all was going well coal consumption tended to be heavy; as soon as the fire started to fragment and release a stream of unburned particles through the tube bank and into the smokebox then consumption could become astronomical without being able to raise steam output above a very modest 30 to 34,000 lb. per hour according to type of coal. Luckily the fireman's strength was not capable single-handed of reaching this volcanic stage, but even within his own capacity, it was clearly approached from time to time in a random manner.

It is easy to understand therefore why L.M.R. inspectors and enginemen preferred their own "Duchess" class on Class 8 duties. Although inferior in cylinder efficiency, it had a rock-like constancy in steam production, and so long as the heating surfaces of its boiler were clean both internally and externally, only the fireman's ability to shovel coal placed any limit to its steam production. The men thus felt at ease that it could support them fully in any extra efforts which the engine might be called upon to make. In writing these lines and in recalling thus what is only a minor incident in the long history of locomotives, I am again filled with sorrow that we were not given the opportunity to eradicate this weakness in 71000. As I have already said, with this one feature rectified, this engine could have been a world beater.

IX

Performance in Service—The Rest

COMPARED TO the Britannias, published records of performances of the other B.R. standard types decline steeply in numbers. The "Clans" rarely seemed to rise above the ordinary in the quality of their performance in such runs as have been published, while the Class 5 4–6–0s, most numerous of the B.R. engines used on passenger duties, were so identical in capacity to the L.M.R. Class 5s that separate records are not numerous; the same comments also apply to the Class 4 2–6–4 tanks. The 2–10–0s intended for freight working, just occasionally blossomed forth as express passenger engines, but these unheralded excursions into high speed running usually caught the faithful recorders napping and hardly any detailed logs remain of runs of which it is reported that speeds up to 90 m.p.h. were attained. Their use on the Somerset and Dorset line very late in the day gave promise of rousing performances, to which impending closure of this line quickly wrote "finis". The other engines of Classes 4, 3 and 2 were not those which, whether of Regional or B.R. form, ever aroused much interest from train timers, for their passenger running was usually associated with semi-fast or stopping trains of easy timing. However, some running, not without interest, is gathered together in this chapter illustrative of what the "rest" were able to do.

Although I had the interesting experience of riding No. 72001, when new out of Crewe, on a Euston to Birmingham express, the ten "Clan" class engines were allocated to Scotland, and spent all their lives there, notwithstanding numerous forays south of the border, mainly with Liverpool and Manchester through-trains and sometimes over the Midland route to Leeds. Records of their running are indeed few and far between. Allen reported in 1952 a run from Symington to Carlisle, set out in Table XXIX. By 1961 when he was writing his book on Pacific locomotives, he had collected no better run with which to typify this particular engine class. This run was recorded from the footplate by the same R. I. Nelson who rode 71000 in Column 1 of Table XXVIII, and this time No. 72002 "Clan Campbell" was the steed. The train was the 1.40 p.m. from Glasgow, with a moderate load from Symington of 11 coaches, 345 tons tare and 360 tons gross. From Symington to Carlisle the train was timed at 68 minutes start to stop for the 66.9 miles, including Beattock Summit 17 miles out. From this latter point there were some tight point to point timings, for example 27 minutes for the 31.1 miles from Beattock to Gretna pass to pass. The hard slogging ascent out of Glasgow to the summit had already been largely completed but drawing away from Symington the driver started with 70 per cent cut-off and partial regulator giving 125 lb. in the steam chest out of the 205 lb. in the boiler; by the end of the first mile the cut-off had come back to 25 per cent and the regulator had been opened to the full, giving 215 lb. in the steam chest out of the full 225 lb. to which the boiler pressure had now been risen. By Lamington cut-off was 15 per cent the speed having now risen to 61½ m.p.h. Full regulator was maintained throughout to Beattock Summit, cut-off being advanced to 20 per cent at Wandelmill, and varied frequently between the limits of 20 and 27 per cent on to Elvanfoot. Boiler pressure was down to 200 lb. before Crawford, but recovered to 225 lb. on the steepening of the gradual incline up the Clyde valley which reaches 1 in 152 to milepost 53½ where the speed had dropped to 49 m.p.h. Then followed a recovery to 58½ m.p.h. on the level 1½ miles past Elvanfoot. Up the 2 miles of 1 in 99 to the summit cut-off was advanced rapidly to a maximum of 34 per cent while the speed dropped to 41½ m.p.h. and the pressure to 190 lb. All this time, with full regulator, the steam chest pressure had kept consistently at 10 lb. below the pressure in the boiler. To this point there had been a loss of ¾ minute on schedule.

Down the other side of Beattock Summit a full 10 minutes was taken for the 10 miles to the station with regulator closed. By Greskine, boiler pressure was 180 lb., evidently deliberate to avoid blowing off, for the fireman seemed to have little difficulty in restoring the pressure to the

Table XXIX
Scottish Region—Symington-Carlisle

Distance	*Engine Load Tare/Full*	*72002 "Clan Campbell" 345/360 Tons*		
		Sched.	*Time*	*Speed*
miles			*m. s.*	
0·0	Symington	0	0 00	—
3·7	Lamington	—	6 17	61½
6·5	Wandemill	—	9 20	55½
9·1	Abingdon	—	12 07	55/57½
11·6	Crawford	—	14 56	56/49
14·3	Elvanfoot	—	18 06	55½/58
17·2	Summit	21	21 42	41½
21·5	Gerskine	—	26 09	67
27·2	Beattock	31	31 36	69
32·4	Wamphray	—	36 22	72/68½
38·2	Nethercleugh	—	41 31	72½
41·1	Lockerbie	43	44 07	67
44·2	Castlemilk	—	46 59	61
46·8	Ecclefechan	—	49 20	73
50·2	Kirtlebridge	—	52 07	76½/66
53·9	Kirkpatrick	—	55 17	77½
58·3	Gretna	58	59 03	72
60·8	Floriston	—	61 08	75½
62·8	Rockcliffe	—	62 51	67
64·9	Kingmoor	—	64 49	66½
66·9	Carlisle	68	67 57	—

56 Class 4 2–6–0 No. 76009 leaving Southampton Central in 1953 on a Portsmouth—Salisbury train.
[P. Ransome-Wallis

57 Class 3 2–6–0 No. 77015 crossing Glenbuck Loch en route to Muirkirk with a train from Carstairs in the early sixties. *[D. Cross*

58 Class 3 2–6–2T leaving Tebay in May 1955 with a miners' special returning from Ulverston to Darlington. *[J. E. Wilkinson*

59 Class 5 4–6–0 No. 73137 with Caprotti valve gear leaving St. Pancras for Leicester in 1958. *[P. Ransome-Wallis*

60 Class 5 4–6–0 No. 73106 nearing Slochd summit en route to Inverness in September 1957.
[W. J. Anderson

61 Class 2 2-6-2T No. 84024 leaving the Portsmouth main line at Peasmarsh Junction with the 18 34 Guildford—Cranleigh train in June 1961. *[R. S. Greenwood*

62 Class 2 2–6–0 No. 78054 leaving Grantown-on-Spey with the afternoon train from Craigellachie to Boat-of-Garten, in September 1957.
[W. J. Anderson

63 Class 2 2–6–0 No. 78009 bound for Leamington Spa soon after departure from Stratford-upon-Avon with a morning local in 1953.
[T. E. Williams

64 Class 4 2–6–4T No. 80003 departing from Glasgow Central in 1955 with an Uplawmoor suburban train.

[N. Faulkner

65 Class 4 4–6–0 No. 75002 passing Thingley Junction, Chippenham, in August 1951, running in on a morning Swindon—Bristol train.

[G. J. Jefferson

66

Canvas draught screens for B.R. tender engines.

67

Class 7 replacement spring hanger bracket, supporting the bottom edge of the main frame in the locality of the welded-in horn guides.

68

Crack in a Class 7 main frame (left) with a temporary repair (right).

225 lb. mark after each spell of firing. At the bottom of the bank full regulator working was restored, first at 15 per cent cut-off, but thereafter advanced to 27 per cent by degrees for the slight rise past Lockerbie. The engine was thus being worked fairly hard to maintain 72½ m.p.h. to Netherclengh and 61 up the final 1 in 200 at Castlemilk; moreover schedule was exceeded by 1 minute to Lockerbie. In the final stages, the regulator was partly closed, but even with only 100 lb. in the steam chest, 76-77½ m.p.h. was sustained past Quintinshill. Notwithstanding another 75½ m.p.h. maximum at Floriston, the engine was still 1 minute down at that point, but a smart finish recovered it, and eventually time was kept exactly to Carlisle.

A snippit was also recorded in the same article with the above, where the same engine showed some smart running between Perth and Stirling as indicated in Table XXX. Speed was soon recovered after the signal check at Forteviot, and at Dunning there begins the steep climb to Gleneagles first at 1 in 121 for 2½ miles, and then at 1 in 100 or slightly steeper. Average speed from Dunning to Whitemoss Box, all on the 121 was 48.9 m.p.h. and up to Auchterarder almost entirely on the 1 in 100 it was as high as 46.1 m.p.h., thus permitting arrival at Gleneagles in 21 minutes 45 seconds actual, 21 minutes net. It was noted that the engine started to blow off immediately the regulator was shut, so that it had by no means been winded by the climb.

Out of Gleneagles, the engine had to start up the last ½ mile of the 1 in 100, but this start was made without any slipping, and there was then some lively travelling from Blackford with an average of 67.2 m.p.h. to Greenloaning, of 75.1 m.p.h. to Kinbuck and of 72 m.p.h. from there to Dunblane. Speed was reduced thereafter, and Stirling was reached in 19 minutes 6 seconds against the 22 minutes allowed. In this way "Clan Campbell" regained, net, 6 minutes of the late start from Perth in this

Table XXX
Scottish Region—Perth-Stirling

Distance	Engine Load Tare/Full	72002 "Clan Campbell" 333/360 Tons		
		Sched.	Time	Speed
miles			m. s.	
0·0	Perth	0	0 00	—
2·0	Hilton Jn.	—	4 53	—
6·8	Forteviot	—	9 59	60
9·6	Dunning	—	13 38*	—
13·7	Auchterarder	—	18 53	46
15·8	Gleneagles	24	21 45	—
0·0		0	0 00	—
2·2	Blackford	—	4 45	—
9·6	Kinbuck	—	11 03	75
12·3	Dunblane	—	13 18	72
17·2	Stirling	22	19 06	—

* *P.W. Check*

relatively short distance. The commentator remarks that the climbing of the engine, from Dunning to Gleneagles was certainly impressive in view of its relatively moderate dimensions.

At this point we come to a rather interesting aspect of locomotive running, for although these engines were, as the above runs show, obviously capable of performance appropriate to their dimensions, they gained a poor reputation amongst enthusiastic train timers. On the other hand the Scottish Motive Power Department and its engine crews had a lively respect for the steaming and running powers of the "Clans" their grumbles being confined to matters such as cab draughts, the height and angle of the tender shovelling plate, and the rather unhandy brake ejector controls, matters which did not affect the running in the slightest degree. To drivers brought up in the rough school of Caledonian and North British Locomotives, even the hard riding quality of these engines passed without comment, while the cab layout received much commendation. Through the kindness of K. R. M. Cameron, Running and Maintenance Officer on the Scottish Region, I have been supplied with the records of a number of runs made by "Clan" class engines in both directions between Glasgow and Carlisle, under motive power observation in the early days when they were first put to work in the region. These records bring out forcefully the difference between what a professional locomotive inspector calls a good run, and what an enthusiastic lay train timer calls a good run, a difference in viewpoint which largely accounts for why the outside observer thought less of the engines than did the regional motive power department.

The footplate inspector reporting back to his regional motive power head office is only incidentally interested in maximum speeds on the easy stretches, and minima on the principal banks. He does not even record these values. On the other hand he is very interested in working to booked overall and intermediate times (to the nearest minute) and to the making up of small amounts of lost time. When an officially observed run is on, the presence of signal checks due to traffic vagaries are reduced to a minimum and only p.w. slacks remain as operating hazards. The presence of the regional H.Q. inspector on the footplate restrains the men from attempting arrivals before time, and from wild bursts of speed which delight the enthusiast but which are usually unnecessary when timekeeping is meticulous. So the run with the least traffic delay which lies closest to the working timetable is the best to our inspector who couples this with other important observations not available to passengers in the train, such as steadiness of boiler pressure and water level in the boiler coupled with economical handling of the engine, and an "off the top" assessment of coal consumption by counting the shovels of coal used, and of water consumption by reading the tender tank water gauge.

Such runs encountered by our timing enthusiast would be rated pedestrian to a degree, and if regularly repeated would earn the class an indifferent reputation. On the other hand the same running, consistently repeated, would be the motive power officers ideal of how a railway should be properly run. The point in common between these two otherwise diverse viewpoints, is that delays due to sloppy operating, and unwillingness of the men to make up a reasonable amount of time if the engine and the coal are right, are signs of inefficiency which are anathema to both. For performances such as those in Table XXXI the "Clans" earned full marks from their owning region; the same performances would probably contain hardly anything worth tabulating to interest the readers of the *Railway Magazine* or the former *Trains Illustrated.*

Viewing the four runs of the table that in Column 2 with No. 72004 was clearly almost a copy book performance with the boiler conditions remaining well nigh perfect throughout. Of the two northbound runs with No. 72003, both meticulous in their timekeeping, that in Column 4 exhibited better steaming for which variation in quality of coal, or expertise on the part of the fireman, or both factors, could have been responsible. Column 1, however, posed a conundrum which clearly baffled the inspector who tabulated the run. Whereas the firemen on the other three runs fired from 277 to 345 shovels-full, the stalwart on 72000, fired no less than 524. Since the water consumption on this run was of the same order as that on the others, the coal consumption was likely to have been pro rata since all the engines were in good condition. The only explanation is that this particular man scooped up a smaller shovel-full each time than did his mates. The inspector tried to allow for this by debiting him only 9 lb. instead of 10 lb. per shovel, but even this, giving 46 lb. per mile was clearly too high, for the engine was not using above 35 lb. of coal per mile based upon other evidence. Coal consumptions such as these must have been very acceptable to men who remembered burning 70, 80 and even 90 lb. per mile with no greater loads on the former Caledonian engines.

To conclude this review of the performance potential of the "Clans" two further observations may be made. The first is that the engines were originally turned out at the end of 1951 with 5¼-in. diameter blastpipes, and from the experience on the test plant with the Class 5 4–6–0 engine referred to in Chapter VII it was clear that this could only provide draught for less output than that of which the boiler was comfortably capable. A sharpening up of the blast by reducing the orifice to 5 in. would obviously provide an improved margin against adverse conditions, and all of the engines were gradually modified in this way during 1953 and 1954, thus making them even more reliable steamers.

A second point is that although having 19½-in. diameter cylinders and 27,520 lb. of starting tractive effort against 19-in. diameter cylinders and 26,120 lb. tractive effort on the Class 5 4–6–0, the Clans had over a ton less adhesive weight, 56 tons 18 cwt. against 58 tons 1 cwt. respectively. The L.M.S. Class 6 three-cylinder "Jubilee" type with which the Clans were directly comparable in power classification, actually had 60 tons of adhesive weight. The Pacifics were, therefore, at some disadvantage in this respect, even on the basis of the diagram book figures. Nobody, however, has reliably recorded the changes in weight on individual axles while running, and there is at least a suspicion, to put it at the very lowest,

Table XXXI

Runs Officially Recorded by Motive Power Inspectors—Glasgow-Carlisle Route

1				2					3				4			
72000 286 Tons to Symington 356 Tons to Carlisle				*72004 280 Tons to Symington 365 Tons to Carlisle*				*Engine* ← → *Load (Tare)*	*72003 376 Tons to Carstairs 301 Tons to Glasgow*				*72003 381 Tons to Carstairs 257 Tons to Glasgow*			
Booked Time	*Actual Time*	*Boiler Pressure lbs./sq. in.*	*Water Level in Boiler*	*Booked Time*	*Actual Time*	*Boiler Pressure lbs./sq. in.*	*Boiler Water Level*		*Booked Time*	*Actual Time*	*Boiler Pressure lbs./sq. in.*	*Boiler Water Level*	*Booked Time*	*Actual Time*	*Boiler Pressure lbs./sq. in.*	*Boiler Water Level*
p.m.				p.m.					p.m.				p.m.			
1·45	1·45	225	F	1·45	1·45	225	F	Glasgow	7·52	7·54*	—	—	7·52	7·52	—	—
1·48	1·48	220	0·75	1·48	1·48	220	F	Eglinton St.	7·49	7·52	220	0·75	7·49	7·49	215	F
1·52	1·52½	200	0·75	1·52	1·52	225	F	Rutherglen	7·45	7·45	220	0·75	7·45	7·45	215	0·75
1·56	1·56	210	0·75	1·56	1·56	225	0·75	Newton	7·42	7·42	210	F	7·42	7·42	215	0·75
1·59	1·59	220	0·75	1·59	1·59	225	0·75	Uddington	7·39	7·39	200	F	7·39	7·38	225	F
2·05	2·04	225	0·75	2·05	2·05	225	0·75	Motherwell	7·33	7·32	225	F	7·33	7·32	225	0·75
2·07	2·07	225	F	2·07	2·07	225	F		7·27	7·26	200	0·75	7·27	7·26	225	F
2·19	2·19	210	0·75	2·19	2·19	225	F	Law Jn.	7·18	7·18	200	0·75	7·18	7·18	225	F
2·31	2·31	220	0·50	2·31	2·33	225	F	Cleghorn	7·10	7·10	220	0·75	7·10	7·09	225	F
2·35	2·35	225	0·75	2·35	2·37	225	F	Carstairs	7·04	7·00	225	F	7·04	7·04	225	F
2·40	2·40	225	F	2·40	2·40	225	F		6·58	6·58	200	F	6·58	6·56	225	F
2·50	2·50	225	0·75	2·50	2·50	225	F	Symington	6·50	6·50	220	F	6·50	6·48	225	0·75
2·55	2·55	225	F	2·55	2·54	225	F									
3·16‡	3·18	225	0·75	3·16	3·17	225	0·75	Summit	6·34†	6·32	210	0·75	6·34†	6·32	225	0·75
3·26§	3·38	220	F	3·26	3·27	225	F	Beattock	6·10	6·10	225	F	6·10	6·10	225	F
									6·06	6·06	220	0·75	6·06	6·05	225	0·75
3·38	3·40	200	0·75	3·38	3·40	225	F	Lockerbie	5·51	5·51	220	0·75	5·51	5·49	225	F
3·53	3·56	220	0·75	3·53	3·55	225	0·75	Gretna	5·31	5·31	225	0·75	5·31	5·29	220	0·75
4·01	4·03	225	F	4·01	4·02	200	F	Carlisle No. 3	5·22	5·22	200	F	5·22	5·22	225	F
4·03	4·05	—	—	4·03	4·04	—	—	Carlisle	5·20	5·20	220	F	5·20	5·20	225	F
	524				277			Coal Consumption no. of Shovels		345				286		
	46·0				27·1			▼ Est. lbs./mile		33·8				33·4		
	28·0				24·1			★Water Gallons/Mile		29·4				28·6		

†Banked up Beattock

that uncompensated "Pacifics" tend to be less surefooted than 4–6–0s other things being equal, a matter to which I return again later in this chapter. So over the hilly routes on which these engines did most of their work, drivers might sometimes experience an instinctive need of caution in handling the regulator which could lead to more pedestrian performances. From all of the foregoing it can by no means be taken that these engines were a total loss, and the fact that no more were built was due to causes other than any connected with their performance or economy. Another small point is that although originally visualised for use on the Perth to Inverness road, they never actually worked there. Cameron tells me, however, that this was not due to any shortcoming of the locomotives, but only because an ample stud of lately built L.M.S. Class 5s was already allocated to this line, and there was no justification therefore, in fitting the "Clans" with tablet exchange apparatus which they would need in order to run north of Perth.

It would be difficult now to assemble a full record of the working of the Class 9 2–10–0s on passenger trains. It was a comparatively rare event for the various species of regional 2–8–0s to find themselves on passenger trains, excepting only the Somerset & Dorset engines which were such a godsend over that mountainous route on summer Saturdays, and except also for the use of L.M.S. 2–8–0s on heavy passenger trains between Glasgow and Aberdeen during the last war. I have seen no published record of this latter circumstance, but the late R. F. Harvey who was in charge of motive power in Scotland at the time used to tell me about it. However, it was not very long after their first appearance that the B.R. 2–10–0s began to appear on relief and excursion workings from time to time over quite widely scattered routes. Who the first hero was, who discovered that these engines would run very freely and steadily at 80 m.p.h. and above, is not known but an inkling of what was to come was described in the *Railway Magazine* for March 1956 in which reference was made to an article in the December 14th, 1955, issue of *The Motor*. It appears that in the course of a demonstration run with an "Alvis" car from Coventry to Glasgow, the driver and his passenger found themselves racing a train through Nithsdale, where road and rail ran side by side for many miles. To quote in part "we began to overtake coach by coach a 14-coach express hauled by a big black brute of a locomotive with numerous massive driving wheels. Soon both driver and fireman had spotted us, and were leaning out of the cab window and waving us on, a challenge which we met by urging the speedometer needle up into the '80s'." The car passenger took a photograph of this exciting contest, and what did the print disclose? Not a "Duchess" or a "Britannia" but No. 92023, one of the Crosti-boilered 2–10–0s apparently making a rendezvous in connection with forthcoming controlled road tests.

In December 1958 Dr. Tuplin was able to disclose that in the previous July, No. 92164 had taken the nine-coach "Master Cutler" over the 23.4 miles of the old Great Central line from Leicester to Nottingham in 23.2 minutes start to stop. Over the 17.5 undulating miles (with a downhill trend) from Rothley to Arkwright St. (Nottingham) the average was 77 m.p.h. with a maximum of 86 at the bottom of the 1 in 176 drop from Baruston Box to the level near Ruddington. Tuplin further reported that a maximum of 90 m.p.h. had been attained by No. 92184 in coming down from Stoke to Essendine on the Grantham to Peterborough section of the old Great Northern main line.

As might well be imagined these runs had caused a bit of a flurry at Headquarters and the various people who had been involved in these speed exploits were instructed to desist.

All was not over, however, for on June 27th, 28th and 30th, and again on July 1st and 15th, 1960, No. 92220 *Evening Star*, worked the "Red Dragon" and "Capitals United" South Wales expresses from Cardiff to Paddington and back. The exact circumstances are shrouded in mystery and are likely to remain so while any of the participants are still on the pay-roll of British Railways. It does seem, however, that *Evening Star* with copper-capped chimney and green paint was something of a pet at Canton shed, and had been used since building on a good deal of passenger working, albeit at moderate speeds. Apparently this engine was standing on June 27th, prepared to work the 10.30 a.m. Cardiff to Portsmouth cross-country train, when the "Britannia" due to work the Paddington train failed. Driver E. Broom in charge of the latter and a senior Canton man, then Chairman of the Local Departmental Committee, was willing to take over the 2–10–0. In his report after the round trip to London he said that an excellent ride was experienced in both directions. The load was 13 coaches, 450 tons, and the engine worked at 16 to 17 per cent cut-off over the level sections. He was of the opinion that a maximum speed of 80-85 m.p.h. was attained running towards the Severn Tunnel on the return journey, although this type of locomotive was not fitted with a speedometer. The consumption of coal and water was stated to be less than average in comparison with "Britannias" and "Castles".

Four more trips were run before officialdom caught up with what was afoot, and then there were bad times for my friend Charles Read in charge of the Newport and Cardiff district and his staff but they all managed to survive. This is a fascinating little corner of locomotive history which it is greatly to be hoped someone some day will expand into a full story.

Of the more legitimate exploits of these engines, Nock records a run made on July 14th, 1961, with No. 92000, one of four 2–10–0s stationed then at Bath Midland Shed for working over the Somerset & Dorset route to Bournemouth. The train was the southbound "Pines Express" and 12 coaches weighing 416 tons tare, 450 tons full were taken through singlehanded in a storm of wind and rain. The well-known and widely experienced French enthusiast, Baron Vuillet was on the footplate, and a very detailed record of the run was set out in the *Railway Magazine* for October 1961. The engine was by now equipped with a double blastpipe and chimney, but the coal was not of the best, consisting of a form of briquettes known as "ovoids". No speed exploits are possible on this route, interest lying in the sustained hill climbing. Over successive grades of the order of 1 in 50 cut-offs up to 48 per cent were used with full boiler pressure remaining rock steady, and the water level in the boiler maintained at the top of the glass. On the ascent from Radstock to Chilcompton average speed over four miles of 1 in 52½ compensated for curvature was 27.7 m.p.h., and both Vuillet and Nock agreed that the average indicated h.p. must have been 2,240, corresponding to an equivalent drawbar horsepower of 2,000. In this case the engine seems to have exceeded the maximum performance as indicated in the test bulletins where at 28 m.p.h. and 50 per cent cut-off only 1,750 equivalent d.b.h.p. was recorded using steam at around 30,000 lb. per hour. In this latter case, however, the engine only had a single chimney and the conditions were sustained rather than transitory. Down at Blandford, 26.4 miles from Bath a rough check of water level showed 3,500 gal. of water to have been used, giving an approximate overall evaporation of 23,000 lb./hr. On the more level sections 60 m.p.h. was attained at which the Baron reported that the engine rode as smoothly as a coach. There was a further mighty effort where after passing Bailey Gate at 64 m.p.h. speed was eased to 50 m.p.h. for the mechanical tablet exchange at Corfe Mullen Junction. On the two miles of 1 in 80 immediately following the reverser was put into 45 per cent cut-off and the regulator opened to the full. A speed of 43 m.p.h. was sustained up this gradient involving estimated cylinder horsepower of 2,700. For this short spell steam was entering the cylinders at a rate far beyond the continuous steaming capacity of the boiler, which once again indicates the punch which the steam locomotive could pack by drawing on the heat storage reservoir in fuel bed and water. Even this was not the end, however, for beyond Poole, on the Parkstone-Branksome bank, 52 per cent cut-off enabled the engine to storm 1½ miles of 1 in 60 gradient at no lower speed than 41 m.p.h. from an initial speed of 47 m.p.h. Baron Vuillet is reported by Nock as summing up this run as "a magnificent example of skilled driving and firing, with high sustained capacity, and remarkable outbursts of exceptional power showing the flexibility of the steam locomotive. This is one of the most remarkable trips I have made on a locomotive, and the Class 9 engines are certainly amongst the very best". Potential such as the above, even although this run was made with skilled and enthusiastic engine men and skilled supervision, indicates that these engines might have operated most successfully on the Highland main line, but the diesel overtook them before the transfer could be made.

The Class 5 4–6–0 engines were spread far and wide over the country, and like their L.M.R. counterparts were seen from Fort William to Bournemouth and from Cardiff to Norwich. With cylinder design and valve events identical with those of the Britannias and 2–10–0s there was no technical limitation at the front end against their delivering comparable performances, subject to suitable adjustment in loading. Although they seem in fact never to have been driven regularly at the same high speeds as the 4–6–2s even when running over the same galloping grounds as the latter, they nevertheless put up quite considerable speed efforts from time to time.

During temporary absence of allocated "Pacifics" a few Class 5s tried their hand at the Liverpool Street-Norwich runs for a brief period, but the men quickly sensed the reduction in both steam production and pulling power compared to their favourite Class 7s. However in Table, XXXII No. 73000 acquitted herself well in a run recorded by Allen in 1952 in which the load was seven bogies, 240 tons gross. Although this load was admittedly light, it was equal to that of the pre-war "East Anglian" then worked by L.N.E.R. "Sandringhams", while the run now described was on a booking 3 minutes faster from Ipswich. No. 73000 passed Manningtree punctually, but a minute was lost on the sharp 8 minutes allowance from there to Colchester. From there to Shenfield 2 minutes was gained on schedule with a time of 29 minutes 6 seconds for the 31.5 miles including some substantial gradients against the engine. Of special note was the rise from 74 to 76 m.p.h. on the level between Kelvedon and Witham, a maximum of 67 m.p.h. on the rising grades after Chelmsford and the speed of 61 m.p.h. maintained up the bank as far as Shenfield. Checks and a stop ruined the conclusion of the run, but the net time was no more than 75 minutes.

Over that other haunt of fast Britannias, from Leicester to St. Pancras, No. 73073 did well on one occasion to gain 7 minutes on the mile-a-minute timing of 99 minutes with a load of 330 tons gross as set out in Column 2 of Table XXIV. In an uninterrupted run no special fireworks were called for with a maximum speed of 83 m.p.h. at Sharnbrook and a minimum of 61 at Leagrave. The average speed for the 88.5 miles from

Table XXXII
Eastern Region—Ipswich-Liverpool St.

Distance	*Engine Load Tare/Full*	*73000 228/240 Tons*		
		Sched.	*Time*	*Speed*
miles			*m. s.*	
0·0	Ipswich	0	0 00	—
3·7	Milepost 65	—	7 07	39
5·5	Bentley	8	9 07	65
9·2	Manningtree	13	12 57*	40/44
12·7	Ardleigh	—	17 26	45/43
14·6	Parsons Heath	—	19 36	70
17·0	Colchester	21	21 51	44
22·1	Mark's Tey	—	27 46	—
26·5	Kelvedon	—	31 29	74
30·1	Witham	34	34 24	76
32·8	Hatfield Peverell	—	36 47	63
36·6	New Hall	—	39 53	72
39·0	Chelmsford	42	41 57	64
45·1	Ingatestone	—	47 44	67/60
48·5	Shenffield	52	50 57	64/61
53·7	Harold Wood	—	57 54*	30/58
58·7	Chadwell Heath	62	62 42	68
64·7	Stratford	69	70 59*	—
68·7	Liverpool St.	77	81 37†	—

**P.W. Check †Signal Stop*

Wigston to Hendon was 68.2 m.p.h. That the "Fives" were not, however, without their own capacity for fast running was exemplified by another run in the opposite direction recorded by Allen in 1954 which on an easier timing of 108 minutes to Leicester nevertheless ran the 19.7 miles from Luton to Bedford North in 14 minutes 57 seconds at an average speed of 79.1 m.p.h., with maxima of 88½ and 92 at Harlington and Ampthill respectively. Minimum after four miles up at 1 in 119 to Sharnbrook was 54½ m.p.h. and speed did not drop below 50 m.p.h. after the climb of 8½ miles from the seventieth milepost, three miles of it at 1 in 118/136 and another three miles at 1 in 200. The load was nine coaches, 320 tons gross.

In another part of the country some lively running is exhibited in Tables XXXIII and XXXIV, in which runs, from Glasgow to Perth, and from Perth to Aberdeen are set out. On the first of these Allen reported No. 73120 with a load of 315 tons gross on the 10.15 a.m. "Granite City" express booked 78 minutes non-stop to Perth. Before Stirling there had been a signal stop near Cumbernauld and a p.w. slack beyond Larbert, but some fine work was done from Stirling up to Kinbuck. It began with a rapid acceleration from 45 m.p.h. through Stirling in two miles but little easier than level to 64 before the engine struck the foot of the bank. Some 2½ miles at 1 in 100 lowered the speed to 51 m.p.h. at Dunblane, while 2¼ miles thereafter at 1 in 88 were climbed at a minimum of 40 m.p.h., the average speed up the 1 in 88 working out at 43.2 m.p.h. The time for the whole 7.6 miles from Stirling to Kinbuck was no more than 8 minutes 53 seconds. The versatility of these engines was then shown by an average of 82.4 m.p.h. over the 11.9 miles from Gleneagles down to Forgandenny with a maximum of 88 m.p.h. and Perth was reached 4½ minutes early. The net time was 69 minutes, however, 9 minutes inside schedule. Allen concludes his commentary with the words "Personally I should have hardly dreamed it possible that with 315 tons an engine of this type could run the 33 miles from passing Stirling over Kinbuck Summit to the Perth stop in 31 minutes 8 seconds but this is what No. 73120 actually did".

The trip north to Aberdeen in Table XXIV with 270 tons gross is one of the very few which have been recorded with the Caprotti version of these engines, and the run was reported by Nock in 1964 at a time when A4 Pacifics had taken over much of the running over the old Caledonian route. Between Stanley Junction and Forfar Nock said he could not recall ever having seen anything quite so speedy with its average of 81.2 m.p.h. over the 15.5 miles from Cargill to Glamis, on a road that is generally level. Further very fast running followed on to Stonehaven with a maximum of 86½ m.p.h. down Farnell Road bank, and a minimum of 60 up the three miles, mostly at 1 in 100 past Marykirk. The average of 73 m.p.h. maintained over 31.6 miles from Guthrie to Dunnotar would also seem to be something of a record, and Stonehaven was reached 6½ minutes early. The train was signalled away after a stop of only 2¼ minutes and Aberdeen was reached by this engine which had run through from Glasgow in 174½ minutes from the starting point, 5½ minutes early.

Table XXXIII
Scottish Region—Glasgow–Perth

Distance Miles	Run No.	1			2			3			Distance Miles
	Engine Load Tare/Full	73120 297/315 Tons			73149§ 238/250 Tons			73149§ 236/245 Tons			
		Schd.	Time m. s.	Speed	Schd.	Time m. s.	Speed	Schd.	Time m. s.	Speed	
0·0	Glasgow	—	0 00	—	—	—	—	—	—	—	—
1·0	St. Rollox	—	3 48	—	—	—	—	—	—	—	—
4·5	Stepps	—	9 50	40	—	—	—	—	—	—	—
6·0	Garnkirk	—	11 22	62	—	—	—	—	—	—	—
9·0	Glenboig	—	14 18†	55	—	—	—	—	—	—	—
13·1	Cumbernauld	—	23 28	—	—	—	—	—	—	—	—
18·5	Greenhill	—	28 12	68	—	—	—	—	—	—	—
22·1	Larbert	—	33 07*	45	—	—	—	—	—	—	—
26·1	Plean	—	37 53	52/72	—	—	—	—	—	—	—
30·2	Stirling	—	41 52	45	0	0 00	—	0	0 00	—	0·0
33·1	Bridge of Allan	—	45 00	64	—	4 11	58	—	4 20	56	2·9
35·1	Dunblane	—	47 18	51/40	8	6 39	47	8	6 50	46	4·9
37·8	Kinbuck	—	50 45	46	—	10 25	41	—	10 14	49	7·6
41·0	Greenloaning	—	54 08	71	—	14 03	60	—	13 24	68	10·8
45·2	Blackford	—	57 57	61	—	18 04	68/60	—	16 58	75/65	15·0
47·4	Gleneagles	—	60 04	—	21	20 13	78	21	18 50	73	17·2
49·5	Auchterarder	—	61 44	84	—	21 48	86	—	20 20	89	19·3
53·6	Dunning	—	63 37	88	—	24 40	90	—	23 15	84	23·4
59·3	Forgandenny	—	68 42	80	—	28 41	87	—	27 32	78/80	29·1
61·2	Hilton Jn.	—	70 22	—	33	30 08	—	33	29 08	—	31·1
63·2	Perth	78	73 30‡	—	37	33 38	—	37	32 35	—	33·0

*P.W. Slack †Signal Stop §Caprotti Valve Gear ‡Net-Time 69 mins.

The Scottish drivers were undoubtedly enthusiastic about these "Caprotti" engines and it is only north of the border that they were really allowed to show what they could do. In England, coming late in the total delivery of B.R. Class 5s they were allocated to sheds and rosters of secondary importance, to such places as Llandudno Junction, Patricroft and Leicester (Midland). I often used to see them when I lived at Radlett on quite secondary trains, and they seldom to my knowledge found themselves on any really fast timings over the St. Pancras route.

As this was being written two further runs by these engines were published in Nock's article in the *Railway Magazine* and are shown in Table XXXIII, each coming very late in the day of steam traction, the engines being used in substitution for Class A4 "Pacifics" on the three-

Table XXXIV

Scottish Region—Perth-Aberdeen

Distance	*Engine Load Tare/Full*	*73153 255/270 Tons*	
		Time	*Speed*
miles		*m. s.*	*m.p.h.*
0·0	Perth	0 00	—
1·6	Almond Valley Jc.	*	—
4·2	Luncarty	—	—
7·2	Stanley Jc.	11 04	60
11·3	Cargil	14 37	57½
15·8	Coupar Angus	18 11	80/76
18·3	Ardler	19 55	83½
20·5	Alyth Jc.	21 35	88/83
24·6	Eassie	24 29	85/82
26·8	Glamis	26 04	88
32·5	Forfar	30 58	82
0·0		0 00	—
2·4	Clocksbridge	4 17	—
7·0	Guthrie	9 12	74
9·0	Glasterlaw	10 42	76/70
12·3	Farnell Road	13 03	86½
15·4	Bridge of Dun	15 22	75/85
18·1	Dubton	17 28	73
19·3	Kinnaber Jc.	18 25	69
21·4	Craigo	20 14	74
32·5	Marykirk	22 06	60 *(min.)*
26·7	Laurencekirk	25 12	70
30·0	Fordoun	27 49	80½
34·0	Drumlithie	31 08	60 *(min.)*
38·6	Dunnottar Boy	35 28	75
41·2	Stonehaven	38 04	—
0·0		0 00	
2·6	M.P. 227½	5 18	33½
4·5	Muchalls	7 38	67
7·9	Portlethen	10 47	61½ *(min.)*
11·3	Cove Bay	13 47	74
11·4	Craiginches S.	16 46*	—
16·1	Aberdeen	22 48*	—

**Signal or P.W. Check*

hour Glasgow to Aberdeen service. In the enthusiastic hands of a St. Rollox crew engine 73149 put up a couple of remarkable performances between Stirling and Perth which showed very vigorous work on the banks as well as much capacity for high-speed running. These runs shown in Columns 2 and 3 gained 3½ and 4½ minutes respectively on schedule over this short heavily-graded section, with maxima touching 90 m.p.h. in each case running downhill towards Perth.

As I have already said, the work of the remaining B.R. engines has not often been recorded, and four further runs must suffice to round off this account of their exploits. The Class 4 2–6–4 tanks, like the Class 5 4–6–0s have been scattered widely over the regions, their principal concentrations being on the L.M. London surburban lines, on the former L.T. and S. line to Southend before electrification, on the Southern Region, and in Scotland. In the latter venue, the run in Table XXXV was recorded from the footplate of engine 80030 by the ubiquitous Mr. R. I. Nelson. The train was the 1.5 p.m. from Glasgow St. Enoch to Ayr, stopping first at Troon. 35.1 miles booked 44 minutes. The train of eight non-corridor coaches weighed 226 tons tare, 245 tons gross. The boiler proved to be free-steaming and pressure was kept at 220 to 225 lb. without difficulty. The driver started with 55 per cent cut-off, dropped to 35 by Cumberland Street and to 25 per cent at the end of the first two miles, increased to 35 in recovering from Paisley slack, and then dropped back to 25 by Elderslegh. From the dead stand for signals between Lochside and Beith the speed capacity of the engine was demonstarted in remarkable fashion.

Table XXXV
Scottish Region—Glasgow-Troon

Distance	Engine Load Tare/Full	80030 226/245 Tons		
		Sched.	*Time*	*Speed*
miles			*m. s.*	
0·0	Glasgow	0	0 00	—
1·6	Shields Road	—	4 20	38½
3·1	Ibrox	—	6 20	48
4·8	Hillington East	—	8 15	55½
6·6	Arkleston Jc.	11	10 18	35
7·7	Paisley	13	11 43	46½
9·7	Elderslie	16	14 07	58½
10·9	Johnstone	—	15 21	61
13·8	How Wood	—	18 29*	50
16·5	Lochside	—	22 05*	42
18·6	Beith	25	28 04†	50½
21·5	Brownhill Jc.	—	31 16	59½
23·3	Dalry	—	32 52	65½
25·8	Dalgarvan	—	35 02	71½
26·8	Kilwinning	33	35 51	74
29·1	Bogside	—	37 39	78½
30·2	Irvine	37	38 28	81
33·8	Barassie	—	45 02†	—
35·1	Troon	44	50 42†	—

**P.W. or Signal Check †Signal Stop*

Less than two miles after restarting, the driver had linked up to 25 per cent, for five miles the regulator was open far enough to give 180 lb./sq. in. in the steam chest, then was eased to give 165 lb. for the next 2½ miles, and then to give only 145 lb. for 3½ miles. For 6½ miles from the stop the line is virtually level; then comes a down grade averaging 1 in 450. Yet with such easy working as this, No. 80030 was doing 50 m.p.h. in less than two miles from the start (Beith), 59½ at Brownhill Junction, 65½ at Dalry, 74 at Kilwinning and 76 at Byrehill Junction, a little less than 11 miles from the start, while the first regulator valve from there onwards produced 81 m.p.h. at the foot of the 1 in 450 through Irvine. Nemesis then followed in two signal stops and a check which made the train 6¾ minutes late into Troon. On the unobstructed lengths the engine had gained all but 2 minutes from Glasgow out to Elderslie, and 1¾ minutes more between Beith and Irvine. Without checks it may be assumed that the engine would have reached Troon 35.1 miles from Glasgow in 38¼ minutes net, nearly 6 minutes inside schedule.

The tender counterpart of the above engines, the Class 4 4–6–0s, were less numerous and less scattered, their main haunts being on L.M. outer suburban trains and on the old L.Y.R., on the Western lines in Central Wales, and on the Southern Region. Amongst other duties they found themselves working the Manchester to Southport trains, once the haunt of the massive four-cylinder 4–6–0 locomotives of George Hughes. Two such runs are set out in Table XXXVI gleaned from *Modern Railways* of March 1962. The engines concerned were 75018 and 75017 with loads of 300 and 285 tons gross respectively. The start from Wigan is for ½ mile down 1 in 155, then follows an undulating stretch to a mile beyond Appley Bridge through which station a slight speed reduction is made, after this it is on the 1¾ miles down at 1 in 282/120/215 past Parbold that the highest speeds are usually reached; and the remainder is mainly level. Over the 16.4 miles from Wigan to St. Luke's the start to stop schedule is

Table XXXVI

L.M. Region—Wigan-Southport

Distance	*Engine No.* *Load Tare/Full*	*75018* *263/300*		*75017* *262/285*	
		Time	*Speed*	*Time*	*Speed*
miles		*m. s.*	*m.p.h.*	*m. s.*	*m.p.h.*
0·0	Wigan	0 00	—	0 00	—
1·1	Douglas Bank	1 46	—	1 49	40
2·7	Gathurst	4 21	54½/57	4 25	50/58
4·5	Appley Bridge	6 17	55	6 25	52
6·7	Parbold	8 30	68	8 47	64/70
8·2	Hoscar	9 44	75	10 03	68
9·6	Burscough Bridge	11 05	69	11 32	65
11·0	New Lane	12 05	75	12 35	70
13·0	Bescar Lane	13 52	66	14 40	74
14·6	Pool Hey Jc.	15 19*	60	15 51	72
15·7	Blowick	16 50*	15	16 52	65
16·4	Southport (St. Lukes)	19 33*	—	17 58	—

usually a smart 20 minutes. No. 75018 hauling a crowded nine-coach train of non-corridor stock attained 75 m.p.h. at two points and kept time despite concluding checks. Net time was 17¼ minutes. The uninterrupted run of 75017 with a slightly lighter load was not quite so fast with its actual time of just under 18 minutes.

These runs would not of themselves be of sufficient interest for inclusion but for the after shadow which they throw on 40 years of locomotive development. Allen comments "I remember (in an article published exactly 40 years ago) being greatly impressed because our four-cylinder 4–6–0 (a Hughes rebuilt and superheated version) had worked up to 69 m.p.h. beyond Parbold. . . . What a vast difference modern valve setting and front-end arrangements generally have made!"

Last of all in this survey of published performance we come to the Class 4 2–6–0, an engine whose 5 ft. 3 in. diameter coupled wheels destined it primarily for freight work, but which like its 2–10–0 big brother sometimes showed a clean pair of heels on occasional passenger jobs. In this latter role its principal appearances were at the two opposite ends of the country, on the Southern Region and in Scotland. In Table XXXVII is set down the only recorded run with one of these engines which I have been able to find, and that over part of an unaccustomed route, namely

Table XXXVII
Scottish Region—Huntley-Aberdeen

Distance	*Engine* *Load Tare/Full*	*76108* *347/365 Tons*	
		Time	*Speed*
miles		*m. s.*	*m.p.h.*
0·0	Huntley	0 00	—
5·8	Gartly	9 30	47
8·0	Kennethmount	13 47	38
9·7	Wardhouse	15 52	67
13·2	Insch	20 24	—
0·0		0 00	
3·0	Oyne	5 09	68
6·2	Pitcaple	8 07	67/68
7·2	Inveramsay	9 12	53/63
10·7	Inverurie	13 28	—
0·0		0 00	
3·5	Kintore	6 27	54
6·3	Kinaldie	9 31	56½
8·6	Pitmedden	11 46	63
10·7	Dyce Jc.	13 50	57
12·8	Bucksburn	15 58	65
14·3	Woodside	17 26	—
15·5	Kittybrewster	19 59*	—
		0 00	
16·5	School Hill	*	
16·9	Aberdeen	4 32	—

**Signal or P.W. Check*

from Forres and Keith on the former Highland Railway, and through Huntley to Aberdeen on the old Great North of Scotland. No. 76108 had a substantial load of 347 tons tare, 365 tons gross. The road is adverse between Rothiemay and Kennethmont but except for 1¼ miles of 1 in 100 after Gartley the gradients are easy, interspersed with stretches of level. It will be seen that speeds from 65 to 68 m.p.h. were freely attained, and the engine showed itself entirely at home in working these short sharp snippets.

No claim is made in the extracts in this chapter and the last that the performances described were better than could be given by other comparable locomotives of regional design. Indeed many of these runs in their initial setting in the *Railway Magazine* or *Trains Illustrated* were tabulated alongside other performances by regional types which equalled or surpassed them. What I have endeavoured to show is that, capably handled and in good order, the B.R. engines were able to fulfil what was designed into them and was displayed in the course of the performance and efficiency tests, and that on the whole they matched the best that designs on the old individual railways had been able to attain. More than this was not expected nor was it possible in any engineering sense having regard to the heat range and thermic cycle over which all Stephensonian engines worked, B.R. standards and regional types alike.

It will have been judged from the foregoing that the B.R. engines had all the ingredients for free steaming and free running, for hard pulling and economical working which footplate staff could wish for, and yet these engines had the most extraordinary variations in the manner in which they were received on the different regions. Indeed the differences were almost unbelievable. At the one extreme, paeans of praise quickly flowed in to us from L. P. Parker in charge of Motive Power on the Great Eastern section of the Eastern Region where the engines were first put to work, as well as from his district superintendents and inspectors. His drivers and firemen fell to with a will and at once deployed the inbuilt capacity of the Britannias to produce running of a quality which in 1951 was not surpassed in any part of the country under comparable circumstances.

At the other extreme, and with equal promptitude, expressions of dissatisfaction and dislike flowed in from the footplate staff on the Western Region, which sometimes reached an almost extravagant level. Not only were we at headquarters treated to this blast of dislike, but certain West Country stalwarts were prepared to regale anyone else who would listen with their views on the iniquities of the new engines. Cecil J. Allen quotes this from one letter which he received from a Western Region enthusiast who had been listening to these tales of woe:

> "To us Western fans the 'Britannias' are just the biggest 'flop' that has ever been designed, and I am certain that most of the men who have to handle them are wondering whatever has thrust such rubbish on them. I suppose the fact is that until the last war and nationalisation, Western folk never really appreciated what excellent engines they had at their command, as they only *saw* strangers from time to time, and

never had the misfortune to handle them. . . . To be fair, however, one hears that they are popular on certain routes of other regions. This may be so but with what kind of engines are their performances being really compared? . . . It is funny, but true that you rarely see Britannias on any important trains on the Western Region."

Indeed, so out of hand did this local phenomenon become that on July 13th, 1957, Riddles had to summon the senior Western locomotive inspectors to headquarters, and with Bond and myself present, try to find out what was the matter—and there proved to be little of any real substance.

In between these two attitudes lay the expected variety of reactions to something new, considering that it was human beings and not robots who had to handle the engines. Railway history is full of examples of enginemen who could or would make nothing of this or that class of locomotive and classic cases include the refusal of Fletcher's drivers on the old N.E.R. to take to his successor McDonnell's new engines in 1884, which led to the latter officer's resignation; the unwillingness of ex-L.N.W. drivers to get the best out of Midland compounds drafted to the Western Division of the L.M.S. in 1925; and much more recently the failure of the G.E. section drivers to get any sort of results out of the S.R. West Countries when a few of these engines were drafted for trial on to that section. In these and many other cases, the same engines were capable of giving entire satisfaction in other hands so that besides all the aspects of mechanical engineering which contribute to the locomotive, a considerable measure of what may be termed psychological engineering is involved as well.

The driver and fireman, who in this country have never been technically trained as mechanics, were rarely much interested in the niceties of design and construction which were so important to their superiors, but had a strong conservative interest in what they had got used to and a very lively suspicion of anything which forced them into new ways of doing. In a rough job in which noise, dirt, draught and vibration were epidemic, they quickly sensed anything which increased still further their physical discomfort, and they had a kind of sixth sense about locomotive sensibilities which a designer might hope would pass unnoticed in daily service. Like the Russians whose attachment to their political beliefs is often less strong than their love of their own national history and traditions, footplatemen, however much in harmony with their unions, the Labour party, and the nationalisation which arose therefrom, usually had an ever stronger attachment to the esprit de corps of their old railway companies which had been drilled into them by all the local hierarchy from general managers down to inspectors.

Coming from the general to the particular, the good feature on the B.R. tender engines in which a firm continuous footplate was provided from boiler back plate to tender front was quite overshadowed by the considerable draught which this arrangement caused notwithstanding provision of gangway doors. This backdraught in its turn caused coal dust to swirl about so that footplate conditions became dirty. Some improvement was made by fitting canvas screens vertically in the gap between engine and tender as shown in Plate 66 and also by reverting to the conventional fall plate, but these were only partially effective, and probably the situation really called for enclosing the whole backend of the cab with a corridor connection to the tender front as was customary on Canadian engines—anyway the criticism was a valid one. So too was that of hard riding on the Pacific locomotives. It has been explained why in adopting the Bulleid design of trailing truck, laminated had been substituted for coil springs on Classes 6 and 7. Space did not permit inclusion of auxiliary springs, so that there was nothing to cushion the effect of high periodity vibrations in the spring system, which in their turn led to vibration and noise in truck side control gear and in the engine to tender coupling. These engines were splendid riders by way of smooth curving and absence of lateral nosing and sway, but here again such undoubted qualities were overshadowed by the personal discomfort to which the hard vertical riding gave cause. By this time the Class 8 Pacifics came out, this shortcoming in the trailing springing was appreciated, and coil springs were re-introduced as on the "Merchant Navies". Although easing the men's discomfort this change introduced new frequencies of vibration which disturbed the fire at high output in the manner already described. On the other hand the other engines in the range, apart from the Pacifics, were very normal in their spring characteristics and did not give rise to any adverse comment as regards riding. A third matter in which the drivers sensed a change of feel on the Pacifics, although not on the other engines, was in the matter of adhesion. This was obviously so in the case of a line like the Western which had never had any Pacifics. Provided with good factors of adhesion, that is adequate nominal weight on the driving wheels in relation to the starting tractive effort, and having sensitive multiple valve regulators, these engines should have been as sure footed as any other but in the way in which they, and all other British Pacifics for that matter, were designed with each axle sprung individually, they were in fact potentially prone to slipping.

Irregularities in track levels, particularly in complex trackwork adjacent to stations could momentarily poise the engine weight on to bogie and trailing truck and partially unload the drivers. A similar condition could arrive also in more permanent form if as a result of spring changing, weight distribution became altered, putting more on the carrying wheels and subtracting it from the adhesive sector. This latter condition could quite readily be arrived at involuntarily when it is remembered that no weighing machines were available at engine sheds. Thus although with careful handling and favourable circumstances the engines were sure footed up to 50 per cent cut-off and full drawbar tractive effort at speed, the driver could never be quite so confident about blasting ahead regardless, as with an end-coupled engine where nothing could unload the coupled wheels and indeed where the greater the tractive effort the more firmly the coupled end of the engine was pressed down on to the rails.

Apart from the individual problem of high output combustion previously described in respect of the Class 8 locomotive, these were the only matters affecting performance in which there was valid cause for complaint, and in which as designers we could have done better than we did. Had steam continued, further batches of these engines could have been better draughtproofed, and provision of softer springing at the back end of the Pacifics would not have been much of a problem. It is only in improvement in adhesion by coupling up all the locomotive springs in a system of compensation, as in American practice, that real difficulty would have been experienced, for more space and additional weight would have been necessary—both difficult to come by under British Railways Loading Gauge and Bridge Curve restrictions.

Beyond, this, deterioration in performance arose only from the many factors which affect all steam locomotives equally, and not the B.R. engines only. All those who chronicle locomotive performance in print have repeatedly drawn attention to the highly diverse attitude of individual drivers. Making up of lost time has always been voluntary on British Railways and its forebears, and however much successful operation depended upon such action, no management ever had the courage to issue precise instructions on the matter, much less institute rewards as was the case in France. All feared the censure which would inevitably have been forthcoming from inspecting officers had any derailment or other mishap been shown to be the result of obligatory hurry to regain lost time. This factor alone gave a lot of lattitude as to how the men drove their engines, and added to this was lack of any rigid enforcement of particular driving standard. Thus full regulator with early cut-off, or partial regulator at later cut-offs were both entirely feasible on most modern engines, and were adopted by individual drivers according to their own taste. Sometimes a strong character such as L. P. Parker on the Great Eastern section could very nearly achieve some uniformity of practice amongst the men but the point is here again that it was never made obligatory by printed instruction and eventual sanctions. Whether all this should have been the case is outside the scope of this book to argue, but undoubtedly such matters affected the level of performance in a random manner, quite apart from the presence of good or illwill, comfort or discomfort. The run of *Dornoch Firth* described in the last chapter illustrated other finer points affecting quality of running, and premature linking-up, mis-management of fire and waterlevel, and unwillingness to run fast were quite common ingredients in poor running. I have said nothing about sheer incompetence, but in locomotive management as in all else, some men are less equal than others, and who, riding unchaperoned on the footplate, has not occasionally come upon men who just did not care—a minority of course, but there nevertheless. We must not single out enginemen, however, as alone exhibiting such vagaries and one has only to think of the gamut of car driving methods to realise that so long as any judgment in control at all is left to individuals, so variations are bound to remain, short of our being completely automated.

Whatever human conduct on the footplate could do towards variety of performance was amplified a thousand fold by all the permutations and combinations of quality of fuel, servicing of the boiler, and the mechanical condition of the engine which could occur.

On the combustion side in addition to the obvious factor of the kind of fuel, lay the fireman's skill in getting the best out of it and the effects of internal dirt in ashpan, tube plate, tubes or smokebox either building up gradually during the course of a run, or sometimes present even before the run began if serviceing was inadequate. On the steam side there could be the effects of priming and of leakages both external and internal. Finally of course an engine run down mechanically, knocking and vibrating all over could give the crew such an uncomfortable time that only the most heroic would still be inclined to run fast and make up time.

All of these things, necessarily hidden from the observer in the train were added to the human element in producing the very variable performances which any given steam locomotive could produce.

Although British Railways receives twice the revenue from freight than it does from passenger working, no writer has been able so far to dramatise the performance of locomotives on freight trains in a manner to make it digestible by the ordinary reader. Although it contains rather more in driving skill, and probably just as much in drama, in difficulties overcome, and in bringing out the characteristics of the different engines, guards journals hardly scratch the surface of what really went on during given journeys, and there is no other volume of recorded data upon which the student can draw. We have nothing, therefore, with which to supplement recorded test results in an examination of performance potential in freight working.

X

Casualities and Defects

WE NOW approach what is by far the most difficult part of setting out a full account of the life and work of any series of locomotives—namely to do honest justice to its record of reliability in service including every aspect of wear, deterioration and failure which could cause it to deliver less than the expected performance. No locomotive has ever been entirely free from this trinity of detractors, and to minimise their effect and to gloss over difficulties experienced leads to insipid narrative which really deceives no one. On the other hand, to describe in meticulous detail adverse circumstances and the reasons for them, can result in such a catalogue of ills that the reader can well wonder how the locomotives ran at all. In between these two extremes, and if any special interest requires it, events can be so selected as to slant the story in any given direction of partisanship or obloquy.

As every teacher knows, the parent is not always the best judge of the temperament and capabilities of his own children. The professional engineer, too, is so interested in the technical challenge of shortcomings in what he has designed, that he is apt to pay scant attention to what is going well, and to let his mind dwell mainly on means of rectification and improvement. I am only too aware therefore, of the dangers lying ahead, but having no longer any axe to grind, I think it should be possible to convey to the reader a sufficiently objective account of this important phase in the history of the B.R. standard engines. At any rate let us try. The close similarity of all twelve types in both basic and detailed design makes it unnecessary to repeat for each one an account of behaviour which was broadly common to all. I have, therefore, taken the "Britannia" as the arch-type, pointing out particular cases where this was untypical, and have divided the defects into three categories, i.e. major teething troubles, non-recurrent after rectification; sources of continuing deterioration or trouble; random failures not following any consistent pattern. It can be said at the outset that the record was very favourable under all three headings. Nothing deep-seated or fundamentally wrong was disclosed by 15 years of use, and however concerned we were at the time with individual happenings, after knowledge of the pains and tribulations of introducing other forms of traction, makes one look back upon those latter days of steam, before resources began to be diverted away from its proper servicing and maintenance, as almost a golden age of relative professional tranquillity.

There were only four major teething troubles, namely water carry over from the boiler into the cylinders with a number of broken piston heads in consequence, an epidemic of coupled wheels coming loose on their axles with or without bent coupling rods as a result—these two on the Pacifics only. There was also fore and aft oscillation transmitted from the engine to the train by all types and finally there were cases of sticking regulators on the 2–10–0s alone.

The first of these events disclosed itself very promptly, for hardly had No. 70000, fresh from Crewe Works and its initial trial run to Carlisle, started revenue working at its home depot at Stratford, when on the rainy Sunday of February 2nd, 1951, Bond and I, hastily summoned by phone, found ourselves viewing the bits and pieces of a fragmented right-hand piston head and cylinder cover. Twelve days later the left-hand head disintegrated, and it took only further periods of nine and six days respectively before the left-hand back cylinder relief valve blew out and disappeared into the blue, and the cylinder cover on the same side fractured. There could be no possible doubt that all this was due to solid water coming over from the boiler into the cylinders in large quantities. The diagrams on Fig. 26 show how this was caused and how it was remedied. In adopting the Superheater Co's. multiple valve regulator in the smokebox for the Pacific locomotives, we had also adopted the same company's centrifugal steam drier attached to the main steampipe intake in the dome as shown in the upper diagram. The idea of this device was that incoming steam was caused by a system of fixed spiral vanes to swirl round in the upper chamber, so that the disengaged droplets of water thus shaken out by centrifugal action could flow back into the boiler proper through the circular orifice seen in the left-hand bottom corner, leaving only dry steam to pass into the cylinders. That was the intention, and it probably so acted at rates of working where the steam velocity was high. On the other hand, with the water at the top of the glass, or above, it could at low rates of working or while coasting pass back up the outlet orifice, or even surge over the top of the steam intake. The immediate remedy is shown in the second diagram where by means of an extension to the intake chamber, a new dome cover, and a deepening of the top nut of the water gauge, an increase from 11 in. to 1 ft. 4⅝ in. was obtained in the distance between maximum visible water level and the point of steam intake. For subsequent application the arrangement shown in the bottom diagram was finally adopted, eliminating the steam drier altogether. As an additional safeguard, cast steel pistons were immediately substituted for the cast iron type originally fitted, and the trouble was thus finally exorcised. Because it disclosed itself so quickly, very few Pacifics were affected before remedial action was taken, and no other B.R. engines were similarly afflicted.

Much more alarming was the occurrence of loose coupled wheels. In July 1951, engine No. 70014 failed with all such wheels shifted on the axle, and six other engines followed in quick succession with shifted wheels and bent side rods in varying degree. This class of engine had not shown itself

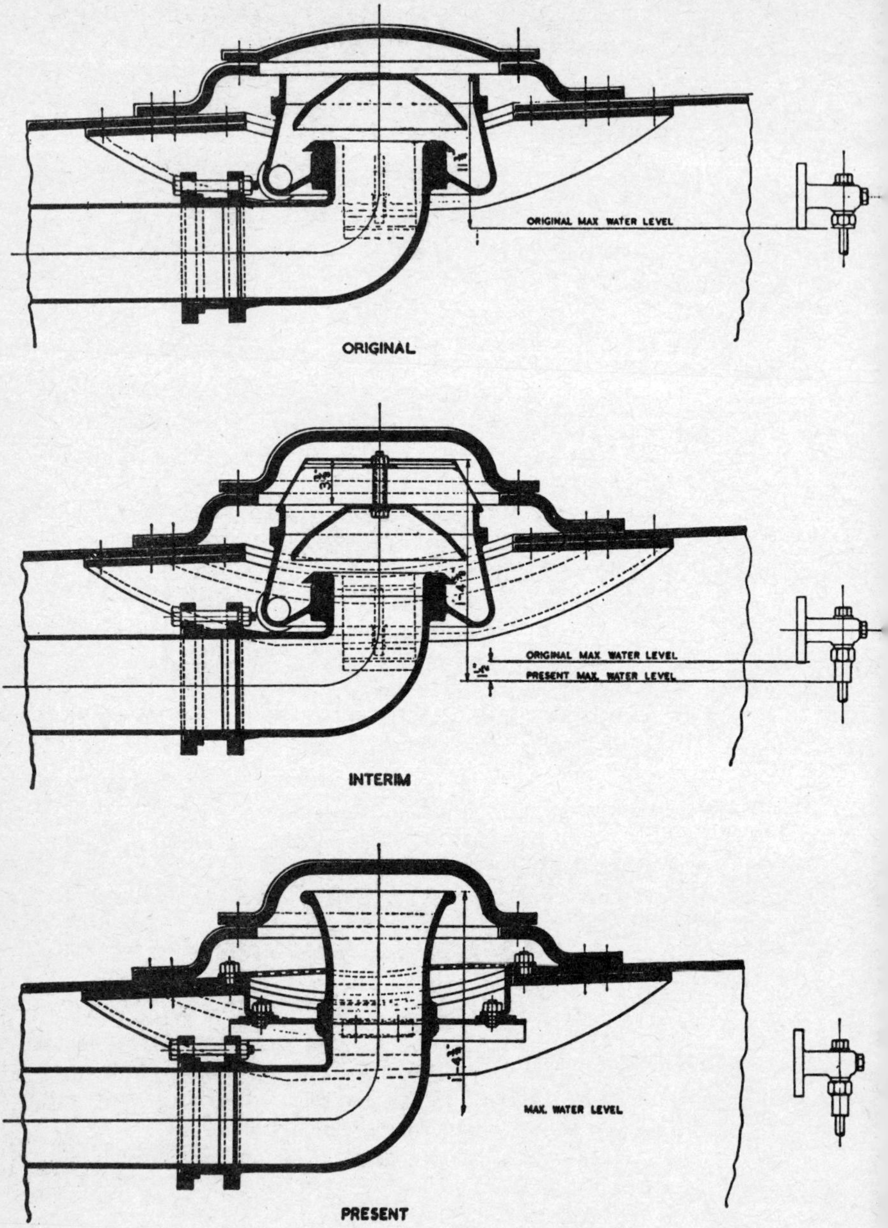

ARRANGEMENT OF STEAM COLLECTOR
B.R. CLASS 7 4-6-2

Fig. 26.

more prone to slipping than other Pacifics generally, rather less so in fact, and the mishaps seemed due to some abnormality in the engines themselves rather than to external circumstances. It was decided to withdraw from service the whole of the 25 "Britannias" so far built until a full diagnosis had been made. It was fortunate that a clue existed in the record of the Class 5 4–6–0 engines to L.M.S. design which had been built with coupled wheel roller bearings between 1947 and 1950. Seven of these had also failed in similar manner over a period of five years, all at low mileages since new. Since these events had been scattered in time, since no further trouble had occurred after rectification, and since 23 other engines similarly fitted had never given any trouble at all, the matter had been allowed to rest.

The above events, together with some evidence on similar lines from abroad, indicated that there were circumstances special to coupled wheels and axles assembled with roller bearings, and that design and workshop practice hitherto quite satifactory for plain bearings might require review. Two differences presented themselves at once:—

(a) Because the roller bearing must pass over the wheel seat in assembly, the latter had to be of smaller diameter than that of the journal on to which the bearing was to be pressed. Thus the gripping surface at the wheel seat was reduced in comparison with the arrangement for plain bearings where the wheel seat was always larger than the journal.

(b) The relatively soft plain bearing with its essential clearances offered some slight cushioning effect to the application of impact forces at the point of reaction. The roller bearing was relatively unyielding in this respect.

A meticulous review of workshop practice was undertaken which disclosed the following:—

(i) The former L.M. practice had been adopted for the B.R. engines in which a 1 in 500 taper on the axle end had been pressed into a parallel wheel seat bore in order to assist an even fit over the whole length of the assembly. This arrangement caused by its geometry a slightly tapered keyway when machined by available methods. Good sideways fit of the key required the keyway to be filed parallel, a laborious process not always carried out.

(ii) Experience with flawed axles on L.M.S. engines had caused key proportions to be altered some years back by the parent office for design in the direction of a shorter, wider and flatter key with resultant reduced side bearing.

(iii) Short cuts had been developed over many years in some workshops alternative to the old laborious but very sound method of hand fitting the keys to within $\frac{5}{8}$ in. of home, final driving in being undertaken by heavy hammer or tup.

Instructions were given for parallel wheel seat fits, deeper and longer keys and the hand fittings of keys to be made standard for all B.R. engines.

Shop equipment in the building works was at that time such that quartering of the crank pin holes (i.e. setting them accurately at the required 90° apart in the case of two-cylinder engines) and balancing of wheel sets could not be carried out after the roller bearings were in position on the axles. It was, therefore, the practice first to press the wheels lightly on to a "dummy" axle. After fitting of tyres, quartering and balancing, the wheels were pressed off the dummy axle, and pressed on to the final axle on which the roller bearings had already been assembled. This double process while not proven harmful, gave added opportunity for marginal errors and has now been eliminated. Lastly, the increased weight of roller bearings had been countered by boring a 4½-in. diameter hole through the axle with a saving of 2 cwts. per axle. Although hollow coupled axles were a commonplace, this particular diameter was at the margin of previous experience, and calculation showed that a slight loss in pressure between wheel boss and axle was possible as a result of squeezing the hollow axle when pressing on. This is an interesting example of the razor's edge which often lies between trouble and no trouble in mechanical engineering matters. Each engine withdrawn, the seven which failed and the 18 still running, had its coupled wheels pressed off, and re-assembly was undertaken having regard to the above findings. A few of the engines had run 50,000 miles without trouble due to the various contributory factors having combined in a more favourable manner, and these engines might have continued trouble free, but it was not possible to take the risk. Among those which showed no sign of trouble was No. 70005, which completed a long series of maximum power tests under conditions which should have shaken the wheels loose in the light of our subsequent findings, but all of the variables must have carried the plus sign on this particular engine. Such an occurrence explains why the building of a single prototype before bulk construction, a practice often recommended, is not necessarily so good a guarantee as is expected against future trouble.

Once wheels can get loose, the shock loading on the side rods which connect them, can greatly increase, and on the well-known principle of belt and braces, a complete survey was made at the same time into coupling rod design. The methods used for design on the former companies were very diverse, not only in the formulae used, but also in the assumptions made, with the result that the same rod seemed to have different margins of safety according to the different methods selected. No absolute procedure for design was ever arrived at, for which a considerable research would have been required to measure all relevant forces and vibrations and to relate them to the elastic properties of the rods—this was one of the several unfinished tasks of the steam era. However, a reasonable compromise between conventional methods of design was arrived at, and replacement coupling rods were produced in which a rectangular cross-section 1⅝-in. wide of depth varying from 5½ in. in the centre to 4½ in. at the ends, superseded a fluted rod 2¼ in. wide on flange and ⅝ in. thick on the web, the depth varying from 5 in. to 4¼ in. In return for an increased weight of 60 lb. on each rod, the estimated lateral deflection at the centre without permanent set was doubled, and the energy absorption at limiting condition was raised by 50 per cent, the quality of the steel remaining the same in each case.

As a result of this comprehensive piece of investigation and the resultant action, trouble of this kind vanished completely. Amongst many who contributed to the work, Gilbert Scholes of Swindon stood out for the painstaking manner in which he took charge of the operation and guided it to a successful conclusion.

The third teething trouble was not, like the other two, lethal, but was the cause of considerable passenger discomfort due to a fore and aft vibration being induced along the train through the couplings. The sufferings of the "hoi polloi" might in this, as in so many other cases, have been overlooked, but when *Iron Duke* and Company began to shake up the supplement paying tycoons on the "Golden Arrow", action became obligatory!

When a two-cylinder engine is balanced so as to reduce vertical hammer blow on the rail to acceptable proportions, a longitudinal component remains which produces a shaking or fore and aft motion. The effect of this can be effectively smothered if the right relationship between unbalanced reciprocating weight and locomotive weight is established mathematically with one important proviso. This is that there should be no artificial "magnifier" along the couplings of the train which will pick up these smothered impulses, and by producing a resonant effect, enlarge them into a perceptible shaking effect once again. It was not until the outbreak of fore and aft vibration with which all the B.R. locomotives were afflicted, whereas their L.M.R. counterparts did not similarly suffer, that this proviso was realised. Previously when this effect had arisen in particular cases, as for example with certain Great Western locomotives, the only remedy which had seemed feasible was to increase the percentage of reciprocating parts balanced to 70 and even in one case to 85 per cent, which was not very popular with the civil engineers.

Now as a result of some intensive work by Sam Ell and his squad at Swindon, assisted by the Brighton team, the real culprit was located in the innocent looking drawbar spring at the rear buffer beam of both tender and tank engines. It would be inappropriate to set out the mathematics here, but suffice it to say that reduction in the initial compression with which these springs were assembled, eliminated a dead portion in the range of drawbar extension under load, extending between a pull of up to 1¼ tons when working under power, and a buffing load of up to 3½ tons when coasting, in which the drawbar spring was not achieving its intended cushioning effect, and the inherent longitudinal vibrations were being transmitted and amplified to the train. With this relatively minor adjustment, complaints died away and the anomalies disappeared. The rogue W.R. two-cylinder engines were found just as amenable to this treatment as were the B.R. engines themselves, an experience which forms an interesting footnote to the theory of locomotive balancing.

The regulator trouble on the 2–10–0s was interesting in that like the fore and aft motion, a previously scattered problem was pinpointed and a final solution was made obligatory. The symptom was that when any considerable degree of slipping took place, sometimes the regulator, which was of the horizontal sliding type, locked in the open position and the driver was unable to shut it. A few such cases had occurred on the former L.M.S. over the years, mainly with Pacific and rebuilt Scot engines. In one particularly dreadful case with the former, the railhead was nearly ground away to the root before by repeated application of the brakes, the driver was able to master the situation and regain control.

It was found that in the process of enlarging the whole steam circuit according to Chapelon's precepts, increase in the size of the regulator ports and therefore of the sliding valve itself had produced a possible condition in which the rapid draining away of steam from the boiler under slipping conditions so reduced the pressure immediately under the valve that boiler pressure on top of it clamped the valve tightly on to its seat and no amount of pulling at the regulator handle could move it until some change in this adverse equilibrium took place of its own accord. The trouble was completely cured by fitting a smaller valve, that from the Class 4, 4–6–0 being substituted. Compared to the Pacific locomotives with their 38 sq. ins. of free area through the multiple valve regulators when fully open, the 2–10–0s now only had 22 sq. in. for the steam to pass through, but this single local constriction in the steam circuit never appeared to produce any ill effects in practice. R. G. Jarvis in charge of the Brighton Office was responsible for this particular solution.

With these exceptional manifestations out of the way we can turn now to the troubles and deteriorations which beset steam traction everywhere—varying in detail of course from design to design, but in sum equally afflicting the B.R. series. This phase is fortunately well documented, for there exist reports from Eastern Region inspectors referring to the good or bad behaviour of no less than 118 different items of B.R. design and equipment, and a Western Region commentary upon this Eastern Region experience. This covers the first 100,000 miles, but through the kindness of D. W. Harvey, Shed Master of the Norwich Motive Power Depot, who has kept the most meticulous records, it is also possible to follow the behaviour of the Britannias throughout the whole 10 years they were working express passenger trains on the Great Eastern section. There have, of course, been other ample and widespread records assembled from time to time, but they have long since passed to the incinerator as being of no current value or interest. However, what remains is fortunately ample for our purpose.

Dealing first with what was good, steaming was lively on all types, free running was excellent, and lateral riding was uniformly steady. Apart from the draught already mentioned, the cab layout was well received and the response of the engines to their controls was alert and ready. Particularly trouble-free and resistant to deterioration were multiple valve regulators, main steam pipe and joints, and all running gear, including axleboxes, slidebars and crossheads. With plain bearings, hot boxes were practically unknown and the wear of the flat surfaces having manganese liners was slight. Roller bearings were trouble-free except for individual cases of defects in the seals permitting entry of water. The valve gear was particularly robust and maintained accurate valve events between shoppings. Tyre wear was moderate and tyre defects negligible, while the tenders as a whole, and tanks and bunkers on tank engines were seldom reported.

On the other hand, cylinder lubrication was erratic and insufficient to avoid fairly rapid piston and valve ring and liner wear. There were all sorts of ingredients in this short-coming, the principal one of which was our old friend, complete inconsistency in its occurrence. Engine classes differed as a whole, while individual engines within a class could depart from the norm in a baffling manner. In the first place there was sheer shortage of oil, but although this could be easily rectified by increasing the flow from the pumps, too much oil led to formation of hard carbon deposits on all the wearing surfaces. As has been described, the engines were at first turned out with regulator controlled atomisers, but most drivers were very averse from coasting with the regulator open even by the small amount required to admit steam to the atomisers. Even on the W.R. which had fathered this arrangement, some of the men would not follow the driving instructions, while on other regions the device was frankly not used to any extent, and by March 1952, it was decided to change over to the arrangement by which supply of steam to the atomisers was coupled to the cylinder cock gear which meant in effect that supply was almost continuous even under coasting conditions. With steam shut off, coasting at speed turns pistons and valves into air pumps tending to suck the contents of the smokebox down the blastpipe, the hot gases scorching the oil and producing the burnt carbon deposit above mentioned. Much variation existed on the former railways as to which was the best coasting cut-off position to minimise this effect. The old practice dating from before superheating of dropping down into full gear was undoubtedly the worst where piston valves were used, although some old stalwarts still seemed to favour it. L.N.E. drivers tended to leave the gear where it happened to be before the regulator was shut, even if this meant less than 20 per cent, while W.R. drivers, and with Stanier's advent, L.M.S. drivers also, had been schooled into the 40–45 per cent position. Later Swindon tests indicated 20–25 per cent as the best, but no position gave immunity, and the so-called air valves which most engines carried, intended to break the vacuum in the steam chests, were ineffective if small, and chattered themselves to pieces if they were large. Another way of reducing the pumping action was to connect the two sides of the piston while coasting by means of bye-pass valves, a feature of many early superheater engines, on which however, the connecting passage was too small to be of the slightest use except at very low speeds. An advantage of the poppet valve where fitted in place of the piston valve was that during coasting without steam the valves dropped off their faces, leaving the full port areas open for passage of air from one side of the piston to the other. Even so at high speeds the resistance to air flow through ports and passages could not entirely avoid suck down the blastpipe. The amount of steam entering through the atomisers was in any case too small to destroy the vacuum, and some railways abroad provided a much larger supply of steam for this purpose separately controlled by the driver, so as once and for all to maintain a pressure rather than a vacuum at the blast-pipe. Such a solution would have proved very hard of acceptance by our drivers however, who unlike their American cousins, had never been used to braking and bringing a train to rest under conditions of steam admission to the cylinders. Yet other factors contributing to the very random behaviour of the lubrication, were first that working with a lower rather than a higher steam chest pressure assisted entry of the oil-bearing atomiser steam. This accounted in some measure for the more consistently well lubricated piston valves which were a feature of Bulleid's Pacifics which were habitually worked on low steam chest pressures. Another facet was that a valve such as the Swindon semi-plug type which retained only the outer rings on each valve head bearing against the liner when steam was shut off, greatly assisted free entry of atomised oil in comparison with the standard six-ring type, which, however, superior in resistance to steam leakage, tended to stifle the entry of oil while the valve heads were passing over the oil entry orifices.

As in the case of the loose wheels, I have dealt at some length with this cylinder lubrication feature, to indicate the kind of problems which still faced us as designers in the last days of steam. The one was satisfactorily solved. The latter was never finally solved, and must be chalked up amongst the other pieces of unfinished business.

It has been explained how any locomotive main frame was a "live" member, subject to a most complex pattern of forces and strains while the engine was running. On the whole B.R. frames gave very little trouble although there were isolated cases as the engines got older, of cracks appearing immediately behind the cylinders where the frame was cut away to clear the bogie wheels. More serious, soon after the engines were turned out, a number of cases of cracks occurred in the unsupported portion of the bottom edge of the main frame between the spring hanger bracket and the horn stay landing as shown on the left-hand side of Plate 68. As a temporary repair, the cracks were cut out and welded, and "straps" were welded across the path of the line of flaw as seen on the right-hand side of the same plate. However, there was an undoubted design weakness here, and when the engines came into shops for their next periodic repairs, and of course for all further new construction, the arrangement shown in Plate 67 was adopted which fully supported the bottom of the frame over the whole of this critical length.

Continuing round the chassis, the single drawbar between engine and tender based upon Doncaster practice, did not prove very satisfactory, and a dramatic case occurred to Nol 70012 which broke its intermediate drawbar near Ilford one day and ran on ahead leaving its tender and train ignominiously behind. Substitution of the L.M. type of drawgear having a solid link coupling, side buffers and additional safety links eliminated further mechanical trouble, and a stopcock was fitted to the brake steampipe leading to the tender, so that in future a locomotive thus careering free would not find itself with hardly any brake power. The reader must remember that destruction of the vacuum due to fracture of the main train pipe when the engine broke away would automatically operate the steam brake valve on the engine, but application of the steam brake itself to the coupled wheels could only be very feeble while steam was escaping full bore through the factured flexible pipe which had supplied the tender.

Piston rod packing of the type consisting of cast iron segments did not prove very good, and renewal was frequent. The very high piston speeds which could be attained by all types was a contributory factor, and the most suitable type of packing for these more onerous conditions was yet to be found, although favourable experience of Southern Region engines with rings of non-ferrous alloys was a pointer.

There was the expected crop of difficulties which afflicted one region more than another, not always readily explainable. For example, tube leakage at the firebox end became quite a widespread trouble at Eastern Region sheds calling for much re-expanding. Western Region depots on the other hand, found no abnormal tendencies in this respect. Although never really run to earth, it seems a fact that Eastern Region maintenance staff never became quite at home with the waterside tube expanders of which both Western and L.M. Regions had had long experience. Again big and little connecting rod ends, although not generally troublesome, gave rather more anxiety to the Eastern than to other regions, again due no doubt to unfamiliarity with the type of white metalling and arrangement of felt lubricating pads proceeding from the Collett-Stanier axis. Such also was the case of the sliding fire-door which W.R. and L.M.R. had always had, and found satisfactory, while other regions complained of a good deal of jamming and difficulty in operation.

A minor but persistent trouble maker was the form of union for small copper pipes which had been developed by Swindon and applied at first to all B.R. engines. In this the pipe end was merely swaged out to form a cone, a most economical procedure compared to the usual practice of brazing a separate cone on to the pipe. However deformation and fracture soon became rife leading to a reversion to the traditional method.

The rocking grates and self-emptying ashpans worked well on the whole, but the former suffered with certain kinds of coal from sticky clinker which sometimes impeded operation, and one of the less attractive tasks for some of the shed staff was to rod out the air orifices in the grate segments to free them from this incubus. Then also, if disposal was mismanaged by tilting the rocking grate to discharge the spent fire into the ashpan without at the same time opening the bottom doors to let it fall straight through into the pit, then the ashpans themselves could warp under the heat and subsequent operation of damper and bottom doors be rendered very imperfect. Ash is one of the most intractable of substances and notwithstanding our great care in the design of the three-compartment ashpan on the Pacifics, accumulations could and did sometimes occur in the back top corners of the outer compartments where the pan was sloped to miss the trailing truck. A possible improvement would have been to fit flushing pipes at this location through which a stream of water could have been directed to dislodge the ash, but this was never undertaken. At the other end of the boiler the self-cleaning smokeboxes worked admirably but a few of the engines required some adjustment to the dimensions of the front horizontal plate before all residual char became automatically disposed of.

As with all motive power, these engines suffered their share of random and minor defects. For the whole life of steam, insignificant and relatively unimportant details have been thorns in the flesh, which continue their teasing even with the newer forms of motive power. There is no superficial reason why nuts should become loose, pipes and other details fracture, or sand cease to flow, and work has been continuous in all design offices to help reduce the appropriate items in the casualty returns. Who for example, would expect a chimney cowl to fall down and smother the blastpipe as happened on one Class 7, and why were individual engines singled out for breakage of smokebox saddle bolts, or bulging of the copper fire-box wrapper plates at the bottom back corners about a foot above the grate. There have been one or two cases of cylinder liners turning round within the steel cylinder castings thus shutting off all entry of oil and exit of condensate through the drain cocks, and random fractures have appeared at piston rod coned ends, return cranks and at one or two points round the Pacific trailing trucks.

Some remaining causes of random trouble were those hardy annuals, with us from the earliest days and never completely exorcised regardless of the generations of engineers who had worked to eliminate them. Of such were burst or leaking superheater elements which while not stopping the job, immediately affected steaming; brick arches becoming displaced or falling down, and coupling rod joint pins subject to excessive wear. There were also mysterious cases of slipping at high speed until it was discovered that coupled axleboxes were sticking in the horns due to the clearances in the manganese liners having been set too fine. And then, irrespective of all that has been said in the foregoing, there was the overriding effect of neglect where for any reason lack of supervision or intolerable staff shortage was present; examination periods unduly prolonged; repairs deferred; tubeplate, tube and brick arch cleaning neglected; all of which took their toll of effective and economical working.

This was the flotsam and jetsam of the casualty returns, obeying no law, following no pattern and rarely justifying design changes. Some could be traced to the human element and suitable cautions administered, but as a whole they had to be lived with, and are still being lived with in this era of new motive power, changed only in the kinds of details which are in use.

In a single chapter no attempt can be made to go into more detail, but what has been outlined overs the gist of the various reports which remain to us, from which a great deal more could be quoted without altering in any material way the impression of the behaviour of these locomotives which it is sought to convey. Of great interest is the similarity of reports from both Eastern and Western Region inspectors. In spite of the initial blast of ill will from the latter region previously mentioned, the Western men were able to report few additional deficiencies over those indicated by their Eastern colleagues and none of a really fundamental nature. Discomfort due to draughts and hard riding still remained as the principal sore spots, coupled with nostalgia for the locomotive and footplate layout which they had been conditioned since their entry into the service to regard as superior to anything else. In May 1953 it was possible for Swindon to write "a great divergence of opinion still exists amongst Western Region enginemen in respect of these engines. Whilst some men now prefer them to regional types, others do not", and with this massive under-statement we must leave the matter.

XI

Mileages and Costs

WE NOW in this penultimate chapter plunge into deep water, and in attempting to round off our consideration of the B.R. locomotives by a reference to the mileages they ran and the repair costs they incurred, we enter a sea of operating conditions, repair philosophy, workshop organisation and financial accounting where much rough sailing and many whirlpools wait to engulf the unwary. Although usually concealed from the eyes of the lay enthusaist, this is one of the most important segments of the whole field of locomotive operation, and our survey of these last days of steam would not be complete without some reference to it. What we are about to deal with are the mileages the engines ran per annum and between repairs on the one hand, and the first cost of the engines and their maintenance cost per mile on the other, matters which at first sight ought to be simple to set out, but, as will be seen, each one must be hedged about by background and proviso's if any real understanding of the figures displayed is to be arrived at. It is the possibility of serious misundertanding if only bald figures are quoted rather than any wish to be secretive which makes railway administrations everywhere very chary about releasing such figures, and particularly cost figures.

The miles which a steam locomotive ran within a day of 24 hours was mainly a traffic matter, and with modern designs, depended very little upon the characteristics of any particular class of engine. If employed on fast trains, and return journeys could be arranged at times convenient to the travelling public, 800 miles was about the maximum possible in this country, while at the other end of the scale, the shunting locomotive, equally performing an essential service, might only travel 36 miles or so during its working day. Grate area and ashpan capcity, together with the provision for coal on the tender, were about the only design features which influenced the maximum values, for a narrow firebox engine might not be able to exceed 300 miles of continuous working, with or without intermediate stops, without having to come off its train for firecleaning and ashpan emptying. Wide firebox engines with good ashpan capacity on the other hand could run 400 miles on one train, as for example from Euston to Glasgow, and for just a very few engines out of the whole stud there was the opportunity for them to be serviced and turned round in time to work another train back to London within the same 24 hours. Such conditions were very rare however, under steam working and 400 miles was considered to be very good utilisation in passenger working.

Suburban and freight service usually offered nothing like this figure due to lower speeds, longer periods for marshalling and layby, or due to absence of traffic demand during appreciable portions of the day.

The annual mileages run were not merely the daily mileage multiplied by 365, but were appreciably less due to the need for examination and repairs at the sheds, and for periodic maintenance in the main workshops, requirements which in 1951 for example required 952 or 5.05 per cent of the total stock of 19,289 to be out of service for workshop repairs and another 10 per cent in round figures to be similarly immobilised at the sheds. Any one locomotive was not, therefore, likely to be available for use by the traffic department on more than around 300 days in the year. Four hundred miles per day for 300 days gives 120,000 miles per annum as the "ideal" practicable utilisation figure but although individual engines have exceeded 100,000 miles per annum if held continuously only on the most favourable rosters, such a mileage was very unusual for a whole class, because some had perforce to work the less favourable turns, and even the very best of engines experienced out of course casualties from time to time. The number of jobs even remotely capable of supporting mileages of this kind were relatively few, and the average annual mileage for each locomotive in stock in the heyday of steam did not exceed a figure round about 26,000 miles. Against this background there is nothing technical to distinguish the more modern locomotives on any of the four regions, or the B.R. locomotives, all of which attained annual mileages appropriate to their duties. "Britannia" 4–6–2 engines for example ran mileages very representative of other Class 7 engines, reaching an annual maximum of around 70,000 miles, and no useful purpose would be served by tabulating the figures throughout the B.R. range.

On the other hand the miles run between consecutive shop repairs is very important as divisor into the overall cost of such repairs, and the whole trend of locomotive design and construction, and the entire repair process has since the grouping of 1923 become more and more concerned in pushing up this mileage as far as it will go compatible with operating reliability in order to bring down to a minimum unit cost per mile run. In the days of inadequate, overloaded, and badly lubricated axleboxes, engines often ran no more than 20,000 to 30,000 miles between visits to the workshops for complete rehabilitation of their running gear, whilst boilers, as senstive to time as to miles in their deterioration would in general run no longer or further than would the engine chassis, and rarely exceeded 50,000 to 60,000 miles between major repairs in which the boiler was lifted off the frames. The steady improvement from this lowly level of performance has been chronicled in numerous articles and technical papers, and the final stage reached in the high noon of steam in the early nineteen fifties has been fully described by R. C. Bond when President Elect of the Institution of Locomotive Engineers.*

* "Organisation and Control of Locomotive Repairs on British Railways", R. C. Bond, *Proc. Inst. Loco. E.* No. 232, 1953.

Table XXXVIII shows the average miles run between consecutive repairs by B.R. locomotives and by the more important regional types in 1957 in which year the former had been coming into service over the previous six years, and there were by then, i.e. by the beginning of that year, 778 of them already in service. To comprehend this table three things require to be understood and first the classification of the repairs themselves. As a locomotive accumulated mileage, the first major components which wore and required rectification were the pistons and valves at from 24,000 to 36,000 miles or less but the cleaning of carbon deposits and renewal of rings could readily be undertaken at the sheds. Eventually wear of tyres either tread or flange predominating, reached an extent beyond which bad riding would be at issue, and axleboxes wore not only on their journal bearing surfaces but on the flat surfaces which rubbed on the horn guides and wheel faces. The renovation of these parts to restore them to the original profiles and clearances was more than the sheds were equipped to undertake, and the whole engine was proposed for and

Table XXXVIII

Repair Mileages—1957

Class of Engine and Operating Region	Between General Repairs		To Intermediate Repair after General		From Intermediate to General or to another Intermediate		Between Boiler Lifts	
	Miles	Time Months	Miles	Time Months	Miles	Time Months	Miles	Time Months
B.R. Cl. 7 L.M.R.	233,797	30·5	136,582	29·5	98,498	19·7	230,733	45·6
E.R.	184,817	27·0	153,158	22·0	138,289	23·0	184,718	27·0
W.R.	224,857	67·5	107,283	32·2	100,698	30·2	220,315	60·6
B.R. Cl. 6 Sc.R.	269,087	62·7			90,218	20·2	269,087	62·7
B.R. Cl. 5 L.M.R.	188,941	57·5	95,052	26·8	55,304	15·2	188,941	57·5
B.R. Cl. 4 4–6–0 W.R.	175,394	68·1	91,231	35·4	83,673	32·5	175,036	66·0
B.R. Cl. 4 2–6–0 S.R.	153,151	51·0	92,111	26·8	78,984	25·2	153,151	51·0
B.R. Cl. 4 2–6–4T Sc.R.	185,995	59·0	97,872	31·3	90,497	28·9	185,995	59·0
S.R.	208,720	63·0	86,551	24·6	—	—	208,720	63·0
L.M. Cl. 8 4–6–2	209,620	34·5	96,650	17·2	91,671	16·2	209,620	34·5
Cl. 7 4–6–0	151,930	28·6	78,615	15·6	61,556	12·6	151,930	29·7
Cl. 5 4–6–0	167,067	50·8	73,014	21·3	64,286	19·2	165,908	49·2
E.R. A. 1 4–6–2	115,259	19·0	108,444	24·0	—	—	115,259	19·0
B. 1 4–6–0	85,382	25·0	—	—	72,346	19·0	82,949	24·0
B. 1 4–6–0 Sc.R.	169,885	53·3	75,917	20·8	65,220	20·8	179,670	52·1
W.R. King 4–6–0	160,717	36·2	88,353	19·9	81,976	18·6	110,751	24·1
Hall 4–6–0	179,396	60·1	88·005	29·5	84,472	28·3	119,960	34·2
S.R. Merchant Navy	158,541	35·8	80,288	15·0	56,075	12·1	158,541	35·8
West Country	206,425	47·2	73,922	18·4	72,953	17·2	200,418	45·0
All Steam Locomotives on British Rlys. 1951	101,130	47	64,268	28	55,183	25	90,871	41

accepted by the main workshops for what was called an intermediate repair. At the same time many other parts on the chassis would have worn or loosened which were also made good, such as rods and valve gear, brakework and countless other details. On modern designs the boiler would at this stage need no more attention than could be given while remaining in position on the frames such as renewal or re-expanding of a number of tubes, and attention to superheater elements, and firebox stays. The amount of work to be done under each heading was decided as a result of initial examination, and the criterion was that after repair the components ought to be able to run a full further period until the next shopping.

Turned out into traffic, the process of wear and deterioration recommenced and at a further period of miles which ought to have been but for a myriad reasons was not always, similar to that run in the first period, the engine fell due yet again for works repairs. In a majority of cases, coincident with this second period the boiler required more serious attention than was possible in position, and it was lifted off the frames and replaced by another of the same kind which had been previously repaired in the boiler and mounting shops. When this occurred the occasion was classified as a general repair. The chassis portion received approximately the same attention as at the intermediate repair, but since still other parts might have become worn, cracked, or loose, there was usually rather more dismantling and rectification to be done at this time. Thus in scanning the table, it will be seen that Columns 2 and 3 roughly add up to Column 1 with a few exceptions shortly to be noted. The last column does not exactly coincide in mileage with the first column for two reasons. The first is that every now and again an out of course boiler defect might arise requiring the boiler to be changed at other than a general repair. The other is that in order to have an economic stock of boilers available for exchanging in the manner described at general repairs, there were more boilers in existence than there were engines, and a given mileage run when divided amongst this greater number, must perforce lead to a somewhat lower unit boiler mileage. Where exceptionally good water was available it was sometimes possible for the boiler to last three periods of chassis repair, so that there would be two intermediate repairs between each general, a condition exemplified by the entry for the B.R. Class 6 4–6–2 solely operating in Scotland where the water is naturally pure, and also by the "West Country" engines on the Southern Region where water was most thoroughly treated by the T.I.A. system.

The second point to observe is that the figures shown are averages, which mask quite wide variations in the repair mileages of individual engines. These variations were due to different standards of maintenance in the running sheds, nature of the services operated, and the quality of boiler feed water as well as due to the good judgment used in workshop examination and the integrity of the repairs carried out at main works. Unexpected design or materials defects could also affect some engines and not others in the same class, thus requiring them to be called into shops prematurely. The random manner in which all of these influences combined, caused the relationship between the different periods and mileages, expressed as averages, to produce figures which did not neatly add up across the sheet, as will also be noticed in the table.

Thirdly, as may well be imagined in respect of a machine subject to so many complex influences, and of workshop practice stemming from so many historical sources under the old companies, more than one idea existed as to the best and most economical procedure for locomotive repair. Just as locomotive design was very different on the four former grouped railways, so after nationalisation, although a common repair procedure was initated by Riddles and Bond, it took some years for even a semblance of uniformity to emerge in the workshops, and vestiges of previous individual practices persisted until the end of steam.

The three main streams of repair outlook were those of the former L.M.S., G.W. and E.N.E. railways. The L.M.S. was the originator and protagonist of undertaking only that work under any given class of repair which the condition of the engine called for under initial examination when it reached the works. The amount of work done was that judged necessary to permit the engine to run a full period until its next classified repair, and nothing was done to those parts in good enough shape to stay the next course. This procedure developed over many years, produced the lowest unit repair cost which was possible within a given engine design and helped availability by reducing the number of days in works to a minimum. On the other hand the examiners, however experienced, were not infallible, and the engines thus repaired did not always run the mileage expected of them, notably the miles run from the intermediate repair to a general was sometimes not so satisfactory as that run from general to intermediate. On the other hand Swindon carried out a most thorough repair every time, and by optical lining up of cylinders, frames and axleboxes and by other means, they could practically guarantee that any engine leaving their works would run the full mileage which its design made possible, whatever the class of repair.

This is seen in the entry for the King and Hall classes in the table. The gross cost of each repair tended to be higher however. It will be noted in passing that boiler change mileages for these Western engines coincide with general repair figures less well than on other regions but knowing the excellence of Swindon boiler practice, this more frequent changing of boilers was more likely to be connected with the quality of the available water.

Doncaster again pursued well into the days of nationalisation its former practice of lifting the boiler of every classified repair as is seen in the entry for the B1 4–6–0. This centre considered that "little and often" was a better repair policy than was the seeking of very high general repair mileages. On the other hand the B1 engines which worked in Scotland came under L.M.S. direction after nationalisation so far as repair policies were concerned, and the quite different figures shown in the second entry in the table for these engines shows how widely these mileages could vary for the same class due only to geographical and administrative location.

Against this background it will be appreciated that merit or the lack of it in repair performance is not to be correctly judged by comparing individual values, and conclusions are only to be drawn from the broad sweep of the figures as a whole. Now the conclusion which the reader is invited to draw is this. The meticulous care which was taken in the design of the B.R. engines in concentrating upon them so many of the developments which contributed to better availability and which have been enumerated in previous chapters, did in fact pay off. The standard engines fully equalled and in some cases were able to exceed the best performances of the regional types. Their concurrence with L.M.S. results was naturally closer than with those of other regions, because the "standards" were developed directly from the former, and there were many common features between the two. This table also pinpoints the different stages in locomotive repair progress in Great Britain. From figures of 30,000 miles between intermediates and 60,000 miles between general of former times, the continual build-up through the years of workshop facilities and practices had by 1947 lifted overall performance to over 50,000 miles between intermediates and 100,000 miles between generals for the whole body of 19,289 locomotives running on British Railways at that time including old and new, and good, bad, and in-indifferent as to design. The more modern design and continuing developments in workshop organisation and equipment was in 1957 producing figures of the order of 170,000 miles between generals, and 70,000 to 90,000 miles intermediately for the more important regional locomotives. The B.R. standards, before the collapse of sustained good maintenance as the diesels took over, were showing signs of a potential of 100,000 miles between intermediates and 200,000 between generals, give or take all the countless extraneous factors which swayed the figures one way and another. This was not because of nationalisation as such or because the engines were to B.R. standard design, but it was the result of all the devoted work over so many years, first in the regions and latterly sponsored by a central headquarters which brought the art of steam locomotive design, construction, and repair to successively higher levels.

During the mileages indicated above the engines lived at motive power depots scattered all over the country. Some were modern, spacious, and well equipped, with means of progressive servicing as the engine passed through the shed, and having highly mechanised coaling and ash handling plants. Examination and repair locations were well lit and well equipped with various aids to the operatives in undertaking their tasks, and such machine tools as were there were reasonably modern and in good order. Some were like this, some provided moderate facilities, but as to others it is better to draw a merciful veil, only wondering in passing how it was possible to expect good maintenance and a willing staff in such dark, dirty, and sometimes even unroofed, accommodation. For in spite of much heroic attention which was bestowed upon individual sheds by successive managements, they remained the cinderellas of the maintenance scene, and as the railways became progressively less profitable, there was perhaps an understandable reluctance to adopt any wholesale renovation or rebuilding.

On the human side the gulf between the best and the worst seemed to be almost as wide, a circumstance bound up not only with the working conditions in the sheds, but with tradition, quality of supervision and the labour situation generally in the district in which the shed was situated. In the latter connection there was an increasing unwillingness on a national scale to undertake work which could be backbreaking and dirty, in the best of conditions, but which in the worst sheds, and with engines no longer properly cleaned, became simply appalling. When the inability of the railway to pay the wages which were obtainable in adjacent industry under much better conditions is considered, and further when it is recalled that staff had to be rostered to cover 24 hours a day, seven days a week, at a time when the five-day week was extending in industry then it is no wonder that hardly the best type of labour could any longer be recuited, and in far too many sheds a small nucleus of experienced old timers, carried upon their overloaded backs platoons of the younger element who hardly cared at all.

Nevertheless there were many places where good relations existed between supervisors and staff and where excellent work was done. It may not be out of place to quote a circular letter issued at Norwich shed on December 20th, 1961, after the B.R. engines had been transferred away and replaced by diesels:—

"To all concerned.

Now that the last Britannia has left Norwich it is fitting that the staff, responsible for their running and maintenance, should know something of their performance in this district during the last decade.

This design, the first of the B.R. standard types, with its greatly increased power and boiler capacity, achieved an immediate success in East Anglia, and made possible the consistent and punctual running of the two-hour "Broadsman" in the then recently introduced hourly interval service of express trains to and from London.

The mileage achieved by the original eight engines since new in 1951 is a measure of the good work put in by servicing and repair staff alike, viz:—

70006	737,716 miles	70010	718,686 miles
70007	690,932 ,,	70011	675,303 ,,
70008	706,852 ,,	70012	717,621 ,,
70009	661,181 ,,	70013	698,458 ,,

During the past summer, as a result of certain modifications to the valve setting, the Britannias have been running through from London on the "Broadsman" without taking water at Ipswich, with a resultant saving of coal, estimated at 80/- per round trip."

Then follows a statement summarising the results of 40 such trips worked by nine sets of men who are mentioned by name, and the letter concludes:—

"... results which reflect greatly on the skill of the drivers and firemen concerned.

In conclusion I wish to express my very sincere thanks to every member of the staff who has contributed to establishing for the Britannias of the G.E. line a reputation second to none.

(signed) D. HARVEY
Depot Master."

What a contrast in encouragement to the staff can be read into this message in comparison with the strict instructions issued at one time by the general manager of another region that no B.R. engine was to be rostered for any train on which he proposed to travel!

To continue as working units in all the circumstances outlined above and to give service over the mileages between visits to the main workshops, which design and repair practices made possible, locomotives simply had to be simple and robust with everything easy to get at, so that inaccessibility could be no excuse for scamping the necessary attention. The B.R. standard engines were specifically designed to meet these requirements, and although there is no statistic which will throw up in isolation how successful they were in this particular environment, there is no doubt that absence of all inside motion, external position of ejectors and injectors and their associated pipe work, mechanical and grease gun lubrication, and ample and accessible facilities for boiler washing out, were very large contributions to keeping up their availability. Even this provision sometimes met with defeat, and where in exceptional cases, neglect had reached an unpardonable stage, it was this very existence of outside pipes and fittings which betrayed leakages all over the engine, and gave substance to those who judged steam as essentially inefficient and decrepit.

Had steam continued, there is no doubt that a very big effort and considerable expenditure would have been necessary, thoroughly to modernise the best sheds and to close up the worst, as one ingredient, but a very important one, towards making more consistent, and then even increasing, the mileage which design made possible between successive workshop repairs. Those who know of, or have read about, the facilities provided at the depots of the Norfolk and Western Railroad in the last days of steam in the U.S.A. will agree, I think that British Railways had still some distance to go in this respect.

If it is necessary to be wary in interpreting figures for locomotive repair mileages, even deeper insight and judgment is required to make anything out of repair costs. All railways tot up the overall cost of their engine repairs and managements make what they can of the ebb and flow of weekly, monthly or annual total costs. It was Sir Josiah Stamp, when President of the L.M.S., who wanted to know where the money was going and who instituted a process of individual costing in order to see how much money was spent on each locomotive. This had an immediate and twofold effect. It greatly improved workshop efficiency because formerly if a man wanted a job of work he had only to take an engine number and charge to "engine repairs" all kinds of extraneous activities. Under individual costing he could make only one charge on any one item for an individual locomotive, because the accountants were now able to keep a strict check on such matters. The second product of individual costing was to present a yardstick for assessing the relative worth of the 393 different classes of locomotive which had existed when the L.M.S. was formed in 1923. In another place* I have described the very wide differences in merit which existed amongst the assorted pre-grouping classes, and this system of costing had no difficulty in turning the limelight on to those types which were inferior under a variety of headings. This costing exercise on the L.M.S. was very through, covering as it did miles run, coal consumption, cost of maintenance in the sheds, and cost of repairs in workshops split up under the main components of which each locomotive was composed. The exercise was also extremely costly from an administrative point of view, as can well be imagined when the large number of bits and pieces requiring examination, repair or replacement at each shopping are considered, each having to be properly recorded and documented. For this reason other railways were not very keen to follow suit, and none of the other three groups introduced anything so comprehensive. The L.M. itself discontinued the exercise when war came, and everything not essential for national survival had to be cut to the bone.

Since the war, and in particular under British Railways, individual costing has only been re-introduced in partial form, and a good deal of controversy has raged within management as to its real value, if any, under more recent conditions. The art of workshop organisation and control has advanced with the years, and adequate internal financial discipline is now obtainable by other and less laborious means than that involved in the former system. Moreover by 1948 the weaker brethren, the really black sheep of the locomotive world, had all been weeded out and it was going to be very much more difficult than was the case in the nineteen twenties to identify marked superiority or inferiority in repair performance in a manner which permitted anyone to do very much about it so far as the remaining engine classes were concerned. Yet again final figures could not be assembled and issued until a considerable period had elapsed since the events to which they referred, and at least three years of such figures were necessary before any meaningful trends emerged. All of these considerations conflicted with the very natural wish still to know what individual engine classes cost, especially as there was now the interest of comparing types from the four former companies amongst themselves and with the new B.R. products. On balance costing was therefore continued for steam in truncated form being confined only to the more important classes, and in respect only of the costs of classified repairs, which meant for the most part costs in main workshops.

At this writing in 1966, it is six years since the last steam locomotive was built, nine years since they were built in any volume and variety. Managerial interest has long since shifted away to the manifold problems of diesel and electric traction. Steam engine costing has ceased and many of the old records have disappeared, no longer justifying the space required for their storage, so that only fragmentary information remains. However,

* "A Modern Locomotive History", E. S. Cox, *Proc. Inst. Loco. E.* No. 190, 1946.

through the kindness of my old chiefs, Bond and Harrison, I have been able to browse through such files of records as remain, tenuously saved from impending destruction, and from these have assembled Table XXXIX which attempts to give a kind of bird's eye view as to how workshop repair costs were looking in the nineteen fifties when steam locomotive repairs were still a going concern. Entries for all of the B.R. standards are displayed, but only for a few of the more important regional types, and the years for which this information is available are spread between 1953 and 1957, over which epoch of inflation there was a rise in labour costs of some 28 per cent.

The column in the table headed "engine" refers only to repair costs of the locomotive itself excluding the boiler, and excluding tender or tanks and bunker. The boiler costs are shown in the adjacent column and this separation was made to solve the problem that a given locomotive did not retain the same boiler throughout its life. Not only were boilers freely changed from one engine to another within any one class, but they

Table XXXIX
Representative Repair Costs per Mile

	Locomotive Class		*Year*	*Engine*	*Boiler*	*Total Loco. and Tender*
				Pence	*Pence*	*Pence*
B.R.	Class 8	4-6-2		10·60	0·95	13·12
	Class 7	4-6-2	1954	7·11	0·88	8·40
			1955	7·13	0·95	8·79
			1956	6·56	0·65	7·89
			1957	6·78	1·09	8·40
	Class 6	4-6-2	1955	6·55	0·79	7·90
			1956	6·38	0·69	7·86
			1957	8·74	0·52	9·99
	Class 5	4-6-0	1954	7·11	0·88	8·40
			1955	5·72	0·68	7·05
			1956	6·91	0·65	8·61
			1957	8·33	0·85	10·19
	Class 4	4-6-0	1956	6·64	0·71	8·56
			1957	7·71	1·26	10·20
	Class 4	2-6-0	1957	6·05	1·51	8·24
	Class 3	2-6-0	1957	6·94	0·83	8·12
	Class 2	2-6-0	1957	5·95	0·94	7·48
	Class 9	2-10-0	1957	9·05	1·53	11·41
	Class 4	2-6-4T	1957	6·96	0·47	7·43
	Class 3	2-6-2T	1957	6·73	0·99	7·72
	Class 2	2-6-2T	1957	8·46	0·43	8·89
L.M.R.	Duchess	4-6-2	1953/55	9·25	2·70	12·70
	Royal Scot	4-6-0	1953/55	8·35	1·30	—
	Class 5	4-6-0	1953/55	6·58	0·74	8·02
E.R.	Class A1	4-6-2	1953/55	6·88	0·94	8·53
W.R.	Castle	4-6-0	1953/55	6·80	2·12	9·73
	Hall	4-6-0	1953/55	5·16	2·33	8·28
S.R.	West Country	4-6-2	1953/55	10·69	0·42	12·33

were often changed from one class to another where a boiler was "standard". There was also the matter of spare boilers, which were carried in stock to facilitate reduction to the lowest level of the days the locomotive need spend in main works under repair, a procedure which might call for the maintenance of, say, 106 boilers to maintain a class of 100 locomotives—this surplus being known as the "economic stock" of boilers.

The accepted accounting procedure was thus to average the repair cost for the whole stock of boilers allocated to a class, including an allowance for the economic stock boilers and for boiler renewals, and these are the costs shown in the "Boiler" column. The last column in the table is the total for the whole locomotive including tender.

Once again if these figures are regarded too narrowly all kinds of anomalies are seen for which there is undoubtedly an explanation, but one which it is beyond the scope of this book to enter into. We are here trying to comprehend the broad trends as they affected the B.R. engines. Just as in mileage run between repairs, so in cost per mile these engines lined up well with the best which modern design, materials, and workmanship could give. As has been mentioned already there has never been any evidence that steam engines in this country cost more to repair as they get older, apart from the change in the value of money, and since by 1957, most of the B.R. engines had attained their rhythm of major periodic replacements such as tyres, the costs shown for the engine portion are reasonably representative of the values at which they were likely to have continued.

A rather striking feature of these chassis costs is the relatively small difference which separates the large from the small engines, on the "per mile" basis. When fitted with modern anti-wear features such as manganese liners on the axleboxes, when good frame design eliminated fractures, and when correct side control limited tyre flange wear, a small engine would run nearly as many miles between repairs as could a large one as Table XXXVIII testifies, which leads to the conclusion that the gross cost of a complete repair was certainly not in direct proportion to size or power. This is perhaps understandable when one considers that in a range of engines in which all except two were two-cylindered and six coupled, the amount of dismantling and putting together again, the man-hours expended on the repair as distinct from the weight of metal to be handled, could not be so very different from one class to another. This is a rather thought provoking consideration when one recalls the many times in history when managements thought they were achieving substantially lower costs by using only the very smallest engines which would handle the traffic.

It is in the column for boiler costs that the widest inconsistencies are to be seen, and this is due to two main factors. Quality of water is the largest single factor in such costs, and variations of 100 per cent and more are possible between consistent use of naturally pure or fully softened water on the one hand imperfectly softened water on the other.

The quality of natural waters varies widely over the country, and the extent and integrity of water treatment varied equally widely from region to region and even between different areas in the same region. Boilers usually, therefore, had to digest mixtures of waters in the course of their travelling about the country the effect of which upon their internal deterioration was very random indeed. This matter of quality of water far transcended the influence of boiler design, although all of the engines shown in the table exemplified good modern practice in this respect. In this context both the B.R. Class 6 in Scotland, and the "West Country" with its painstaking treatment of the water in each individual boiler by the T.I.A. system, were able to exhibit boiler repair costs of the order of only one halfpenny per mile.

The second factor is that the introduction of the heavier items of repair such as retubing or replacement of tubeplates is slower in boilers than are comparable major replacements in the chassis so that it takes longer for the boiler repair costs of a whole locomotive class to level out at their ultimate value, an aspect which leaves the boiler costs of B.R. locomotives in 1957 somewhat lower than they might have averaged out had a ten-year period, say, been able to be recorded.

On the whole this long exercise in individual costing of steam locomotives has been a little disappointing in its practical application as distinct from its interest. If it was attempted to interpret the figures to compare one locomotive class with another, the whole thing was unbalanced by the effects of varying workshop organisation and procedures, of which piece-work prices individually negotiated in each shop, and overhead costs appropriate to the equipment and layout of each works were important ingredients.

On the other hand, if the costing system sought to judge between the different repair proceedures, to compare the results of the more meticulous but more costly Swindon repair as against the less all-embracing but cheaper Crewe repair for example, then the variations in operating conditions, in shed practices and in the design of the locomotives themselves intervened to cloud the issue. Eventually had steam continued, a common design factor would have eliminated one variable, but it was not to be. And yet, when all is said, cost comparisons, broken down sufficiently small to compare particular design, construction and repair features are an obvious component of effective technical management, so it is no wonder that under dieselisation, the problem is still being worked upon by engineers and accountants alike.

Finally, there is the first cost of the locomotives to be considered, which in the form of interest and depreciation charges loads the total running costs throughout the life of the engines. The whole of the B.R. engines were built in railway workshops and the costs set out in Table XL are what are known as departmental costs. This shows what it actually cost the railway to build including a full share of overhead and administrative costs. But it does not include interest on capital laid out during construction, nor does it include anything for a selling organisation or for

Table XL
Representative First Cost (Departmental)

Locomotive Class	*71000*	*70000*	*72000*	*73000*	*75000*	*76000*	*77000*	*78000*	*92000*	*80000*	*82000*	*84000*
	£	£	£	£	£	£	£	£	£	£	£	£
Year 1951		17520		16618	15163					15337		
1952			20426			16892		14377			12995	
1953												12246
1954	33919*	22600					16054		22263			
1955				21293					23853†		14627	
1956								16871				
1957				26381*	19682	20577				20222		16511
1958									28277			

*Caprotti Valve Gear. †Crosti Boilers

makers' profit, items which private manufacturers, but not railways, must include in their final selling price. Throughout railway history, controversy has continued round this point, the railways claiming they could build cheaper than the trade, and manufacturers claiming unfair competition because all possible costs were not included.

A second factor in considering these first costs is to recall the rise in prices of raw materials and of wages which took place in these years resulting in successively higher costs as time went on, even for the same article. Table XLI shows the effect of price variation on these costs year by year from 1948–1964 in respect of steam construction.

Then even in a given year, the identical engine could not be built in different works for the same price. In considering repair costs it was mentioned how piecework prices subject to local agreement, and closely anchored to local history varied for the same process from works to works. Similarly building methods, machine tool equipment, and overhead costs of all kinds—even the very price of gas and electricity—varied about the country, so that it was not necessarily a sign of managerial merit that works A could produce more cheaply than works B, although it was certainly the aim of management to narrow the gap. Here again this gap could be appreciable, a potent factor in deciding which works were to be closed in the great rationalisation of railway workshops which has taken place more recently on British Railways. This said, it is interesting to recall how very cheap steam locomotives were, and in days when £80,000 buys only moderate power and £120,000 spent does not achieve anything like three times the performance potential of a steam engine which would cost £40,000 today, it is a matter for mild surprise that so little attention was seemingly paid to the capital expenditure part of the total cost of the traction ton mile, in the great changeover.

Table XLI

Estimated Percentage Increase on First Cost of Steam Locomotives Due to Price Variation

Year	Percent	Year	Percent	Year	Percent	Year	Percent
1948	171	1952	95	1956	54	1960	36
1949	152	1953	81	1957	46	1961	30
1950	136	1954	74	1958	42	1962	25
1951	118	1955	66	1959	39	1963	19
						1964	9

The figures for any one year, show by what percentage building costs in that year would have to be increased in order to equate them to the price standards of 1965.

XII

Conclusion

IF THE facts set out in the preceeding chapters are allowed to speak for themselves, then no concluding apology is necessary for the B.R. standard steam locomotives. That they were both effective and economical can be confirmed from the many aspects which have been displayed in these pages, and they could have well supported a more extended end to the steam era, had substitution by other forms of motive power been organised at a more sensible and less costly pace. Indeed in two directions they could have enhanced operating potential. At the high-speed express passenger end it required only a little more work on the test plant and in the drawing office to permit successors to No. 71000 to have been turned out capable of exceeding the performance of any express passenger then in use, and at lower rates of fuel consumption. Secondly, the smaller engines in the series had it in them to support considerable accelerations in secondary duties, coupled with a big increase in utilisation, had operating practice been able to break loose from its traditional habits, and undertake as much reorganisation for steam as high first cost has compelled it to introduce for diesel traction. Three out of the 12 types, 72000, 77000 and 82000 would no longer have been required with the progressive removal of civil engineering restrictions, but had the rest continued to have been built a few years longer, these together with the more modern of the former regional types would have formed a body of steam power under whose umbrella electrification schemes could have matured as capital became authorised, and diesel traction could have been introduced only as fast as fully developed, proven, and reliable motive power became available. The money which would have been saved by this policy is incalculable, and all that was needed to make it effective was a mental attitude, a determination starting from managerial level at the top and permeating the whole service, that steam was good until it was superseded by something better, and that until it could be replaced it must continue to have first-class supervision, driving and firing, servicing, and maintenance. One has not had to travel very far into the continent of Europe to see this policy in action during similar periods of gradual change-over in the form of motive power, and its rightness is so self-obvious that it is curious how it has escaped the preception of management in this country.

None of the facts about the B.R. locomotives in this book have been seriously challenged on engineering grounds, but these engines have been under extensive criticism both inside and outside the railway in that they were brought into existence at all. It has been explained how, given the fact of nationalisation, and the personalities who were thrown up at that time, the production of these locomotives was inevitable. But what if there had been no nationalisation? We can perhaps draw this story to a close, a story of steam itself as well as of the particular B.R. standards, by leaving aside hard fact, and by permitting our fancy to roam over what could have happened had the four former main-line companies continued independently until the present day, and then soaring further into the stratosphere, to consider how steam might have responded had some external factor, such for example as very cheap coal, eliminated the diesel competitor, and had compelled steam to serve rail transport needs for a much longer period.

If the old railways had continued, it is not too difficult to imagine what would have happened over the first decade at any rate. On the L.M.S., Ivatt's effective incumbency as C.M.E. would have been prolonged, and almost certainly Bond would have succeeded him. As this line were pioneers in diesel traction, its extension was to be expected in both main line and shunting form. The presence of F. A. Pope within the organisation would have assured re-introduction of the lightweight diesel railcar, and there is no reason to think that electrification from Euston to Liverpool and Manchester and even to Glasgow would not have been pushed forward. It is unlikely in my opinion that under L.M.S. auspices, the diesel substitution would have occurred at so frantic a pace as eventually took place under the B.T.C., so that further improvement of the breed of steam engines would undoubtedly have continued, and the lavish provision of testing equipment on that line would have permitted attack upon the various pieces of unfinished business mentioned in this book which would have led to the further refinement of this kind of motive power in the form of improved design. Whether any new steam types would have been designed is largely a matter of how enterprising and forward looking the operators might have become. Heavy fast freight working might have caused the four-cylinder 4–8–4 locomotive to mature which had been outlined in 1942, whilst if for any reason authority for capital expenditure on main-line electrification had had to be delayed, then the 4–6–4 proposal of later Stanier years might have been dusted off and re-examined. Had firebox and ashpan design proved to be too restricted above the trailing bogie within the existing loading gauge, or had the increased bridge curve loading which these engines demanded been denied, then it is just possible that high power requirements in steam might have veered off towards a thoroughly modernised "Garratt". But if none of this had matured by around 1957 or 1958 or so, then it would never have appeared at all for by then the diesel and the electric would have been ready to take over for all new construction.

On the E.N.E.R., Harrison would probably have succeeded Peppercorn, and the Gresley tradition would undoubtedly have been given a further lease of life. By 1948 this railway was badly in need of modernised motive power in the lower categories, and one if not two medium axle load 2–6–0s would almost certainly have appeared of Doncaster design in place of the Derby designs which were actually introduced. As regards main-line passenger services it was even then likely to be a neck and neck race between steam and diesel, for schemes were maturing, sponsored largely by the Chief Electrical Engineer for a fairly large-scale exercise on the East Coast line with 2,000 h.p. diesel units, while 1,500 volt electrification, already extant on the Shenfield line and coming into service on the Manchester-Sheffield route was a further contestant in the game. Before, however, such schemes could have come to fruition in any widespread manner it would have been surprising if Harrison had not come up with a super A4 Pacific, or had not even had another look at Gresley's own 4–8–2 scheme. A more powerful freight type than the O1 2–8–0 would also possibly have been required.

I cannot think that the above prognostications for the two larger groups would have been very far off the mark, but when we come to the other two railways it is much more difficult to prophesy. On the G.W., Cook would have succeeded Hawksworth, and no break in the Churchward tradition is thus to be imagined. Whether standard Swindon types, including the Castles would have continued to be built in the late 1950's is a moot point, and even mooter is whether, at any point in time, Cook might have broken away in the direction of a Pacific in substitution for the "King". There is no doubt that a 4–6–2 was sketched out in Swindon drawing office in pre-nationalisation days, but apparently at a somewhat lowly level, for Hawkesworth when taxed with its existence is reputed to have returned a very dusty answer. Ell's contribution, with his controlled road testing and the stationary plant, would have become available, but it is only fair to say that he and his boys found more freedom and encouragement under Riddles than they had done under the latter days of pure G.W. rule. How and when the newer forms of motive power would have been taken up is a question mark, for the Western people had started off down a very doubtful road with their two gas turbine locomotives. Inevitably at some point they would have had to realise the unsuitability of this form of prime mover for British traction conditions, and by then there might have been some useful diesel experience elsewhere for them to consider. The diesel hydraulics would no doubt have appeared in the time and quantity of their manifestation on the Western Region of British Railways—perhaps even earlier.

It is on the Southern that the oracle falls silent for Bulleid would have been faced with 36 "Leaders" already approved by his Board, and how he would have tackled the almost complete initial failure of this unorthodox machine is indeed a fascinating conundrum. Extension of third rail electrification might have proceeded somewhat faster under purely S.R. auspices but even so, there would still have been an operating requirement for a lot of medium-powered engines for duties eventually undertaken by L.M. and B.R. 2–6–0s and 2–6–2 and 2–6–4 tank engines. Whether also the three 1–C–C–1 main-line diesel locomotives authorised before nationalisation would have been the first of many, or whether they would have made no headway against the over-supply of West Country and Merchant Navy Pacifics on the one hand, and approaching electrification on the other, is also veiled from us. All that is reasonably certain is that as long as he was in harness, Bulleid's volatile mind would have continued to produce engineering shocks of one kind or another, and that his successor, whoever he might have been, could be assured an interesting, if less dramatic, time of it in getting things back on to an even keel once more.

These are the things which nationalisation possibly prevented us from seeing, but prophesy even short term is an uncertain game, and a single new personality, with other antecedents and background, dropped into a position of power on any of the railways, could have greatly changed the picture which has been outlined. The broad trend away from steam and towards the diesel and electrification would, however, in my view have remained unchanged. All that would have varied would have been the pace of its onset. But suppose by some miraculous circumstances there had been no reasons for a broad trend away from steam, how could the demand for increased power have been dealt with? If effective competition was to be offered to the aeroplane and the motorway, great increase in power would have become necessary and means would have been required to achieve this economically, that is without materially increasing the overall operating cost per horsepower. This the diesel and the electric have in fact done by a combination of higher relative capital costs, and lower unit running costs, but what were the possibilities with steam? This is a fascinating question which would require a book on its own to answer properly, but some indications may be given here in order to round off this account of the last days of steam.

Before doing so a word is necessary about steam locomotive power, as such, which is far from being a finite quantity for any given engine design, as may have been judged from what I have already written. The highest published horsepower figures for any British locomotive are those recorded for the Stanier Pacific No. 6234 on a special dynometer car test run in 1939 from Crewe to Glasgow and back with 600 tons when 3,348 indicated horsepower, and 2,511 h.p. at the drawbar were calculated at particular instants of time. The engine was here making use of that attribute of the steam locomotive where the thermal reservoir capacity of a large boiler could permit a steam user for a limited period greater than that which could be produced at the maximum practicable rate of firing. Table XLII shows these figures in the first entry, associated with an estimated steam user of 48,000 lb./hr. Whereas under constant conditions on the test plant, using best coal, and two firemen, 42,000 lb. of steam was the maximum output which could be sustained for one hour, as shown in entry 2. The corresponding i.h.p. was 2,900, but neither this power nor any above it could be rostered for the locomotive in daily service on the line, because they were far beyond the capacity of a single fireman, they allowed nothing for less than perfect conditions, and were not sustainable for more than an hour or so before various deteriorations, particularly in the state of the fire, would impose a gradual reduction in output. Moreover such rates were not only uneconomical but mechanical reliability could not be safeguarded in continued use at such outputs.

The third entry shows the rating for this engine which was at the same time within the sustained capacity of a single fireman, and also represented the maximum efficiency at which the engine could work. Non-stop runs of up to six hours' duration could be safely rostered at such an output, representing 1,610 i.h.p. and 1,460 h.p. at the drawbar, with short spurts using the reservoir capacity of the boiler into the region of 2,000 i.h.p.

Even these figures were very variable according to the capacity and willingness of the fireman, so an obvious first stage in reaching for higher power would have been to apply mechanical stokers to the largest existing locomotives. As was indicated in Chapter VII when the test results of the stoker fitted 2–10–0s were described, such a fitting did not mean that the maximum outputs of entry 2 in the table could be immediately applied to daily service. In the first place under stoker firing about 16 per cent more coal was used for the same steam output, and assuming 7,000 lb. of coal per hour was at roughly the front end and grate limit for the "Duchess" locomotive, entry 4 in the table estimates the maximum steam production and the horsepowers which might have proved attainable under test plant conditions. Continued working in service at these rates was not, however, a practical proposition, for the reasons outlined above, and entry 5 sets out my personal estimate of what might have proved

Table XLII

Actual and Estimated Performance—L.M.S. "Duchess" Class 4-6-2

Entry No.	*Condition*	*I.H.P.*	*D.B.H.P. on Level*	*Steam Rate lbs./hr.*	*Coal Rate lbs./hr.*	*Coal lbs./sq. ft. Grate per hr.*	*Boiler Efficiency*
1	Instantaneous all-out performance using reservoir capacity of boiler	3348†	2511	48000	—	—	—
2	Maximum output on test plant sustained for one hour	2900	2300	42000	7160	140	58
3	Output at firing rate of 3000 lbs. best coal per hour	1610	1460	23500	3000	60	77
4	Estimated maximum output on test plant with mechanical stoker	2680	2260	38000	7160	140	58
5	Estimated practical maximum sustained output in service with mechanical stoker	2150	1800	30000	5000	100	68

†This was a calculated figure and was probably over-estimated

sustainable in day-to-day working on the line had coal and maintenance of first-class quality been assured. Here again skilful driving and firing could have provided for transient outputs towards 2,500 i.h.p. over short sharp efforts of acceleration or climbing, whilst still remaining within economical conditions of working both thermally and mechanically.

Of course oil firing could have taken the place of the mechanical stokers, but this measure is unlikely to have produced materially greater power. This is because in exchange for losing the various problems of solid fuel combustion and ash disposal, there would have been substituted the difficulties of attaining the optimum mixing of oil fuel and air within the severe limitation in volume of the locomotive firebox.

Consideration of this alternative, however, rather negatives the premise on which any possible continuation of steam could have been founded. If coal was the only fuel available then railway engineers would have had to soldier on with steam development; but the moment oil was present, then it must be in the diesel engine cylinder at over 30 per cent efficiency that it would be used, and not via the boiler in steam engine cylinders at less than 15 per cent efficiency.

The next stage is to visualise undertaking by steam the work of the Deltic diesel locomotive, which is operating the most vigorous schedules under this form of power in 1966. This machine can sustain 3,300 h.p. at the engine output shafts, 2,600 h.p. at the drawbar at constant speed on the level, all day if required, at no detriment to its efficiency or mechanical well being. At 60 m.p.h. it can still maintain 2,500 d.b.h.p. and delivers a drawbar pull at that speed of just on 7 tons. It is clear that, even mechanically stoked, an engine of the size of the "Duchess" or of the B.R. Class 8 could not reach such performance and that an increase in power in the neighbourhood of 40 per cent would have to be sought.

A steam locomotive which might just have reached this level of performance was outlined in Stanier's time in 1938, and was illustrated in diagram form on page 125 of Volume I of *Locomotive Panorama*. This 4–6–4 engine could have exerted a tractive effort of 7.6 tons at the wheel-rim at 60 m.p.h., and although this would have been reduced at the tender drawbar by the pull needed to overcome the resistance of locomotive as a whole, nevertheless the grate area of 70 sq. ft. and the free area through the tube bank available from a boiler diameter of 6 ft. 10 in. should have permitted the desired increased of 40 per cent in steam production and horsepower to give figures under the same conditions as those of entry 5 in the table, but for this engine amounting to 3,000 i.h.p. and 2,500 d.b.h.p.

But everything was now grinding against the limits of the possible in steam construction under B.R. conditions. Axle weights of 24 tons even with no hammer blow could only reluctantly have been conceded by the civil engineer—the boiler diameter was at the absolute limit, and firebox volume in relation to grate area was barely sufficient for best combustion. The arrangements for air supply to the grate and for ash disposal on long runs would have presented tricky problems. Finally a total weight of

187 tons of which only 42 per cent was adhesive compared unfavourably with the Deltic's maximum axle load of 18 tons, and total weight of 106 tons, all of it adhesive. A 4–8–4 version could have improved adhesion and reduced the individual axle load somewhat, but at the expense of unfavourable boiler proportions since only the barrel length and not the free area through the tubes could have been increased. Thus the conventional locomotive as we have known it would have reached its ceiling in this country at about this level of output, and could not have matched the still higher performance which electric traction is already making possible on the Euston-Manchester-Liverpool services today. For such further advance it would have been essential to turn to the Garratt principle, with which it might have proved possible to push up power into the region of even 5,000 i.h.p.

To round off this examination of the outer boundaries of steam capacity suitable for high-speed services, what more has it been found possible to do in the rest of the world, where size and permissible weight did not bear so restrictively on locomotive development? There have been two answers, one achieved in actual operating fact, and the other worked out in considerable detail by a steam engineer of international fame but never translated into actual locomotives. The largest and most modern American locomotives such as the New York Central 4–8–4 and the Pennsylvania 4–4–4–4 types have sustained on test 6,600 i.h.p., 5,000 h.p. at the drawbar as a maximum output, and were rostered in daily service to utilise continuously something of the order of 4,000 i.h.p., 3,200 h.p. at the drawbar, corresponding to a drawbar pull of around 9 tons at 60 m.p.h. But such locomotives were so vast in both dimensions and weight that there was no possibility whatever of their like being introduced into this small island with its severe limitations of loading gauge and bridge curve.

The other answer was that proposed by Andre Chapelon. I do not know how many locomotive enthusiasts have seen this engineer's monumental work published in 1938, *La Locomotive a Vapeur*, the size of a family bible, and running to over 900 pages, close packed with technical discussion and formulae of the most erudite kind. Foreseeing what steam might have to contend with in order to compete with electrification which was already being closely considered by the S.C.N.F. heirarchy, Chapelon included at the very end of his book detailed designs for seven new types of steam locomotive to meet a specification calling for up to 6,000 i.h.p. at the cylinders and maximum speeds up to 125 m.p.h. In the four most powerful of these of 2–12–0, 4–8–4, 4–6–4 and 2–10–4 wheel arrangement, six cylinders were to be employed, two h.p. and four l.p. driving in pairs on to different axles using two crank axles. Chapelon relied upon all the features he had developed for his earlier locomotives to produce 6,000 h.p. out of engine weights of about 160 tons (excluding tender), whereas an American engine of similar power weighed 210 tons. Even although axle loads varied from 22 to 29 tons, much of the design seemed of a very slender nature, with crank axles and coupled wheel bearings looking very near the bone. Remembering the sad fate of his thermally wonderful 4–8–0s whose frames and mechanical parts would not stand up to high power output in service, one wonders whether these new proposals would have fared much better. Nothing was said about mechanical stoking and one wonders what supermen were to be bred in France who could fire at the rates demanded even by a workable proportion of the maximum power. Moreover the ashpans had only been inserted in the most sketchy manner below the grate and above the trailing bogies, and clearly not much thought was given to the effect of ash accumulation on long runs at high sustained effort. So, fascinating although these projects were to the student, there was little about them which was going to be applicable to the British scene if and when steam was required to take a big leap forward in power in that country.

Thus it seems abundantly clear that steam was reaching the end of the road so far as its further development towards higher power was concerned, and that the B.R. standard locomotives, which are the subject matter of this book, could only have had a limited extension towards further designs before performance and efficiency rounded off at a level below that to which diesel, and to a far greater extent electric traction can aspire in facing up to the demands of the future.

To conclude, I can think of no better epitaph than to repeat the concluding sentence of the chapter I previously wrote on the B.R. engines in *Locomotive Panorama*: "I for one feel that the B.R. standards were no discredit to the men who designed and built them, and above all to Robin Riddles who so staunchly upheld them. It could be that posterity will support this opinion when all the heat and dust of current partisan advocacy has died away."

Index

General

Locomotives

Personalities